T0224991

Computer Architecture and Design Methodologies

Series Editors

Anupam Chattopadhyay, Nanyang Technological University, Singapore, Singapore
Soumitra Kumar Nandy, Indian Institute of Science, Bangalore, India
Jürgen Teich, Friedrich-Alexander-Universität Erlangen-Nürnberg (FAU),
Erlangen, Germany
Debdeep Mukhopadhyay, Indian Institute of Technology Kharagpur, Kharagpur,
West Bengal, India

Twilight zone of Moore's law is affecting computer architecture design like never before. The strongest impact on computer architecture is perhaps the move from unicore to multicore architectures, represented by commodity architectures like general purpose graphics processing units (gpgpus). Besides that, deep impact of application-specific constraints from emerging embedded applications is presenting designers with new, energy-efficient architectures like heterogeneous multi-core, accelerator-rich System-on-Chip (SoC). These effects together with the security, reliability, thermal and manufacturability challenges of nanoscale technologies are forcing computing platforms to move towards innovative solutions. Finally, the emergence of technologies beyond conventional charge-based computing has led to a series of radical new architectures and design methodologies.

The aim of this book series is to capture these diverse, emerging architectural innovations as well as the corresponding design methodologies. The scope covers the following.

- Heterogeneous multi-core SoC and their design methodology
- Domain-specific architectures and their design methodology
- Novel technology constraints, such as security, fault-tolerance and their impact on architecture design
- Novel technologies, such as resistive memory, and their impact on architecture design
- Extremely parallel architectures

More information about this series at http://www.springer.com/series/15213

Sujoy Sinha Roy · Ingrid Verbauwhede

Lattice-Based Public-Key Cryptography in Hardware

 Springer

Sujoy Sinha Roy
School of Computer Science
University of Birmingham
Birmingham, UK

Ingrid Verbauwhede
ESAT—COSIC
KU Leuven
Leuven, Belgium

ISSN 2367-3478   ISSN 2367-3486   (electronic)
Computer Architecture and Design Methodologies
ISBN 978-981-32-9996-2   ISBN 978-981-32-9994-8   (eBook)
https://doi.org/10.1007/978-981-32-9994-8

This Springer imprint is published by the registered company Springer Nature Singapore Pte Ltd.
The registered company address is: 152 Beach Road, #21-01/04 Gateway East, Singapore 189721,
Singapore

List of Publications

Journals

1. Sujoy Sinha Roy, Frederik Vercauteren, Jo Vliegen, Ingrid Verbauwhede, "Hardware Assisted Fully Homomorphic Function Evaluation and Encrypted Search", Accepted in IEEE Transactions on Computers as a regular paper. Preprint available on IEEE Xplore, DOI: 10.1109/TC.2017.2686385.
2. Sujoy Sinha Roy, Junfeng Fan, Ingrid Verbauwhede, "Accelerating Scalar Conversion for Koblitz Curve Cryptoprocessors on Hardware Platforms", In IEEE Transactions on Very Large Scale Integration (VLSI) Systems, vol. 23, no. 5, pp. 810–818, May 2015. 2015.
3. Donald Donglong Chen, Nele Mentens, Frederik Vercauteren, Sujoy Sinha Roy, Ray CC Cheung, Derek Pao, Ingrid Verbauwhede, "High-Speed Polynomial Multiplication Architecture for Ring-LWE and SHE Cryptosystems", In IEEE Transactions on Circuits and Systems I: Regular Papers, vol. 62, no. 1, pp. 157–166, Jan. 2015.
4. Oscar Reparaz, Sujoy Sinha Roy, Ruan de Clercq, Frederik Vercauteren, Ingrid Verbauwhede, "Masking ring-LWE", In Journal of Cryptographic Engineering, Springer Berlin Heidelberg, 2016, Vol.6(2), p.139–153.
5. Zhe Liu, Thomas Pöppelmann, Tobias Oder, Hwajeong Seo, Sujoy Sinha Roy, Tim Güneysu, Johann Großschädl, Howon Kim, Ingrid Verbauwhede, "High-Performance Ideal Lattice-Based Cryptography on 8-bit AVR Microcontrollers", Accepted in ACM Transactions on Embedded Computing Systems.
6. Kimmo Järvinen, Sujoy Sinha Roy, Ingrid Verbauwhede, "Arithmetic of τ-adic Expansions for Lightweight Koblitz Curve Cryptography", Under review in Journal of Cryptographic Engineering, Springer Berlin Heidelberg.

Conferences and Workshops

1. Sujoy Sinha Roy, Angshuman Karmakar, Ingrid Verbauwhede, "Ring-LWE: Applications to Cryptography and Their Efficient Realization", In International Conference on Security, Privacy, and Applied Cryptography Engineering, Springer International Publishing, Volume 10076 of the book series Lecture Notes in Computer Science (LNCS).
2. Oscar Reparaz, Ruan de Clercq, Sujoy Sinha Roy, Frederik Vercauteren, Ingrid Verbauwhede, "Additively homomorphic ring-LWE masking", In International Workshop on Post-Quantum Cryptography, Springer International Publishing, Volume 9606 of the book series Lecture Notes in Computer Science (LNCS).
3. Jeroen Bosmans, Sujoy Sinha Roy, Kimmo Järvinen, Ingrid Verbauwhede, "A Tiny Coprocessor for Elliptic Curve Cryptography over the 256-bit NIST Prime Field", In 2016 29th International Conference on VLSI Design and 2016 15th International Conference on Embedded Systems (VLSID).

4. Sujoy Sinha Roy, Kimmo Järvinen, Frederik Vercauteren, Vassil Dimitrov, Ingrid Verbauwhede, "Modular hardware architecture for somewhat homomorphic function evaluation", In International Workshop on Cryptographic Hardware and Embedded Systems CHES 2015, pp 164–184, Springer Berlin Heidelberg, Volume 9293 of the book series Lecture Notes in Computer Science (LNCS).

5. Zhe Liu, Hwajeong Seo, Sujoy Sinha Roy, Johann Großschädl, Howon Kim, Ingrid Verbauwhede, "Efficient Ring-LWE Encryption on 8-Bit AVR Processors", In International Workshop on Cryptographic Hardware and Embedded Systems CHES 2015, pp 663–682, Springer Berlin Heidelberg, Volume 9293 of the book series Lecture Notes in Computer Science (LNCS).

6. Sujoy Sinha Roy, Kimmo Järvinen, Ingrid Verbauwhede, "Lightweight coprocessor for Koblitz curves: 283-bit ECC including scalar conversion with only 4300 gates", In International Workshop on Cryptographic Hardware and Embedded Systems CHES 2015, pp 102–122, Springer Berlin Heidelberg, Volume 9293 of the book series Lecture Notes in Computer Science (LNCS).

7. Oscar Reparaz, Sujoy Sinha Roy, Frederik Vercauteren, Ingrid Verbauwhede, "A masked ring-LWE implementation", In International Workshop on Cryptographic Hardware and Embedded Systems, CHES 2015, pp 683–702, Volume 9293 of the book series Lecture Notes in Computer Science (LNCS).

8. Ruan De Clercq, Sujoy Sinha Roy, Frederik Vercauteren, Ingrid Verbauwhede, "Efficient software implementation of ring-LWE encryption", In DATE '15 Proceedings of the 2015 Design, Automation & Test in Europe Conference & Exhibition Pages 339–344.

9. Ingrid Verbauwhede, Josep Balasch, Sujoy Sinha Roy, Anthony Van Herrewege, "Circuit challenges from cryptography", In 2015 IEEE International Solid-State Circuits Conference - (ISSCC) Digest of Technical Papers, San Francisco, CA, 2015, pp. 1–2.

10. Sujoy Sinha Roy, Frederik Vercauteren, Nele Mentens, Donald Donglong Chen, Ingrid Verbauwhede, "Compact ring- LWE cryptoprocessor", In International Workshop on Cryptographic Hardware and Embedded Systems CHES 2014: pp 371–391, Springer Berlin Heidelberg, Volume 8731 of the book series Lecture Notes in Computer Science (LNCS).

11. Sujoy Sinha Roy, Frederik Vercauteren, Ingrid Verbauwhede, "High precision discrete Gaussian sampling on FPGAs", In International Conference on Selected Areas in Cryptography, SAC 2013 pp 383–401, Springer Berlin Heidelberg, Volume 8282 of the book series Lecture Notes in Computer Science (LNCS).

Contents

Abbreviations

AES	Advanced Encryption Standard
ALU	Arithmetic and Logic Unit
ALU	Arithmetic Logic Unit
ASIC	Application Specific Integrated Circuit
BLISS	Bimodal Lattice Signature Scheme
BRAM	Block RAM
CDT	Cumulative Distribution Table
CRT	Chinese remainder theorem
CVP	Closest Vector Problem
DDG	Discrete Distribution Generating
DES	Data Encryption Standard
DH	Diffie-Hellman
DLP	Discrete Logarithm Problem
DPA	Differential Power Analysis
DRAM	Distributed RAM
DRU	Division and Rounding Unit
DSP	Digital Signal Processor
ECC	Elliptic Curve Cryptography
ECDLP	Elliptic Curve Discrete Logarithm Problem
ECDSA	Elliptic Curve Digital Signature Algorithm
FF	Flip Flop
FFT	Fast Fourier Transform
FHE	Fully Homomorphic Encryption
FPGA	Field Programmable Gate Array
IoT	Internet of Things
LPR	Lindner-Peikert-Regev
LUT	Lookup Table
LWE	Learning-With-Errors
MAC	Multiply and Accumulate
MPC	Multiparty Computation

NTT	Number Theoretic Transform
PALU	Polynomial Arithmetic and Logic Unit
PIR	Private Information Retrieval
PKC	Public Key Cryptography
RFID	Radio Frequency Identification
ring-LWE	Ring-Learning-With-Errors
SHE	Somewhat Homomorphic Encryption
SIMD	Single Instruction Multiple Data
SPA	Simple Power Analysis
SVP	Shortest Vector Problem

List of Figures

List of Tables

Chapter 1
Introduction

Since the advent of the internet, our world has become more and more connected every day. The International Telecommunications Union reports [19] that the number of internet users has increased from 400 million in 2000 to 3.2 billion in 2015. This growth rate is expected to be faster in the future as a result of internet penetration in the developing nations. The Internet of Things (IoT) is a network of connected devices ranging from powerful personal computers and smart phones to low-cost passive RFID tags. These devices are capable of computing together and exchanging information with or without human intervention and are present in many areas of our life such as smart homes, smart grids, intelligent transportation, smart cities. By 2020 there will be 21 billion IoT devices [19]. These connected devices could upload their data or even outsource costly computational tasks to a cloud server. A cloud server is a very powerful device with huge storage and computation capability. Indeed cloud computing and IoT are tightly coupled.

In this connected world, our daily life applications such as email, social networks, e-commerce, online banking and several others generate and process massive amounts of information every day [23]. Snowden's revelation [20] in 2013 has brought security and privacy issues into the spotlight of media coverage. Google, Facebook and other leading internet companies are facing increasing pressures from government spying agencies to reveal information about the users. Now users are more concerned about security and privacy than before. Therefore it is of vital importance to protect digital information by incorporating confidentiality, integrity, and data availability.

Cryptography is the science of protecting digital information. In a broader sense, our present day cryptography schemes can be split into two branches: the symmetric-key cryptography schemes and the public-key cryptography schemes. In a symmetric-key cryptography application, the two communicating parties use a common key to protect their information. The existing symmetric-key cryptography schemes are computationally very fast. However their security is based on the assumption that the two parties agree on a common key secretly before initiating the communication. Public-key cryptography is free from this assumption as there is no need for a common

© Springer Nature Singapore Pte Ltd. 2020   1
S. Sinha Roy and I. Verbauwhede, *Lattice-Based Public-Key Cryptography in Hardware*,
Computer Architecture and Design Methodologies,
https://doi.org/10.1007/978-981-32-9994-8_1

key. This feature makes public-key cryptography schemes very attractive despite their slower performance. In practice, most cryptographic protocols use both symmetric-key cryptography and public-key cryptography in tandem: a public-key cryptography scheme is used to agree on a common key and then a symmetric-key cryptography scheme is used to secure a large amount of digital information. In this research we concentrate only on public-key cryptography.

The most widely used public-key cryptography schemes are the RSA cryptosystem [17] and the elliptic-curve cryptosystem (ECC) [9]. Security of the RSA and elliptic-curve cryptosystems is based on the hardness of the integer factorization problem and elliptic-curve discrete logarithm problem (ECDLP) respectively. Although the RSA cryptosystem is conceptually a simple scheme, the main disadvantages are its large key size and slow private key operations. In comparison, ECC requires much smaller key size as the ECDLP problem is much harder to break than the integer factorization problem [2]. Nevertheless ECC requires expensive computation. To improve the efficiency of ECC, numerous proposals have been published in the literature in the past three decades. Such proposals [7] include choice of finite field, choice of elliptic-curve, optimizations in the finite field arithmetic, design of efficient algorithms, and finally tailoring the algorithms for applications and platforms. Designing ECC for lightweight IoT applications with low resources has been an extremely active research field in recent years [1, 3, 6, 8, 10, 11, 15, 21, 22]. These proposals focus predominantly on 163-bit elliptic-curves which provide medium security level of about 80 bits. However recent advances in the cryptanalysis have brought the 80-bit security level too close to call *insecure* for applications that require long term security. NIST recommended phasing out usage of 160-bit elliptic-curve cryptography by the end of the year 2010 [13]. Recently FPGA-based hardware accelerators have been used to solve a 117.35-bit ECDLP on an elliptic-curve over $\mathbb{F}_{2^{127}}$ [5]. In the first part of this thesis we address this problem by designing a lightweight ECC coprocessor using a high security 283-bit Koblitz curve (Table 1.1).

Public-key cryptography is an ever evolving branch of cryptography. With our present day computers, RSA and ECC schemes are considered secure when the key size is sufficiently large and countermeasures against side channel and fault attacks are enabled. However this situation changes in the domain of quantum computing. In 1994, Shor [18] designed a quantum algorithm that renders the above schemes insecure. In 2018, Google announced a 72-qubit quantum computing chip 'Bristlecone' that could lead to a major breakthrough known as 'quantum supremacy'. However, to break the 1024-bit RSA cryptosystem using Shor's algorithm, a million qubit quantum computer will be needed.

Table 1.1 NIST recommended approximate key length for bit-security [2]

Algorithm	80-bit	112-bit	140-bit
RSA	1024	2048	3072
ECC	160	224	256

Though there is no known powerful quantum computer today, several public and private organizations are trying to build quantum computers due to its potential applications. In 2014 a BBC News article [4] reports that the NSA is building a code cracking quantum computer. Post-quantum cryptography is a branch of cryptography that focuses on the design and analysis of schemes that are secure against quantum computing attacks. Beside the scientific community, several standardization bodies and commercial organizations are considering post-quantum cryptography. In 2016 NIST recommended a gradual shift towards post-quantum cryptography [14] and called for a standardization process for post-quantum public-key cryptography schemes. The results of the round 1 phase of the post-quantum cryptography standardization processes are expected to be announced in 2019.

Amongst several candidates for post-quantum public-key cryptography, lattice-based cryptography appears to be the most promising because of its computational efficiency, strong security assurance, and support for a wide range of applications. Lattice-based cryptography has become a hot research topic in this decade since the introduction of the Learning-With-Errors (LWE) problem in 2009 [16] and its more efficient ring variant, the ring-LWE problem in 2010 [12]. Till 2012 almost all of the literature addressed the theoretical aspects of LWE and ring-LWE-based cryptography and very little was known about the implementation feasibilities and performance aspects. This motivated us to investigate implementation aspects of ring-LWE-based public-key cryptography in this work.

1.1 Summary of the Book

In a broader sense the book investigates implementation aspects of next generation public-key cryptography on hardware platforms. The organization of the book is illustrated in Fig. 1.1.

- **Chapter** 2 We give a brief introduction to the mathematical background of the ring-LWE problem. Then we describe the public-key cryptosystems that we consider for implementation in this research.
- **Chapter** 3 In this chapter we focus on an efficient implementation of a Koblitz curve point multiplier targeting a high security level. Koblitz curves are a class of computationally efficient elliptic-curves that offer fast point multiplications if the scalars are given as specific τ-adic expansions. This needs conversion from integer scalars to equivalent τ-adic expansions. We propose the first lightweight variant of the conversion algorithm and introduce the first lightweight implementation of Koblitz curves that includes the scalar conversion. We also include countermeasures against side-channel attacks making the coprocessor the first lightweight coprocessor for Koblitz curves that includes a set of countermeasures against timing attacks, SPA, DPA and safe-error fault attacks.
- **Chapter** 4 The focus of this chapter is to design a high-precision and computationally efficient discrete Gaussian sampler for lattice-based post-quantum cryp-

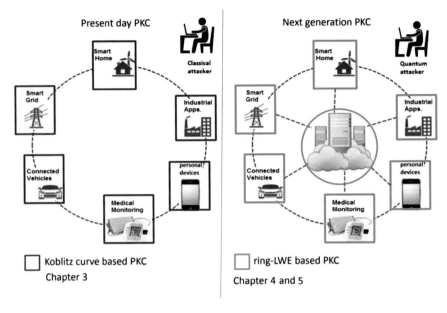

Fig. 1.1 Structure of the book

tography. Discrete Gaussian sampling is an integral part of many lattice-based cryptosystems such as public-key encryption schemes, digital signature schemes and homomorphic encryption schemes. We choose the Knuth-Yao sampling algorithm and propose a novel implementation of the algorithm based on an efficient traversal of the discrete distribution generating (DDG) tree. We investigate various optimization techniques to achieve minimum area and computation time. Next we study timing and power attacks on the Knuth-Yao sampler and propose a random shuffling countermeasure to protect the Gaussian distributed samples against such attacks.

- **Chapter** 5 In this chapter we design an efficient and compact ring-LWE-based public-key encryption processor. The encryption processor is composed of two main components: a polynomial arithmetic unit and a discrete Gaussian sampler. For the discrete Gaussian sampling, we use the Knuth-Yao sampler from Chap. 4. For polynomial multiplication, we apply the Number Theoretic Transform (NTT). We propose three optimizations for the NTT to speed up computation and reduce resource requirement. Finally, at the system level, we also propose an optimization of the ring-LWE encryption that reduces the number of NTT operations. We use these computational optimizations along with several architectural optimizations to design an instruction-set ring-LWE public-key encryption processor on FPGA platforms.
- **Chapter** 6 This chapter concludes the thesis and formulates future works.

References

1. Azarderakhsh R, Järvinen KU, Mozaffari-Kermani M (2014) Efficient algorithm and architecture for elliptic curve cryptography for extremely constrained secure applications. IEEE Trans Circuits Syst I-Regul Pap 61(4):1144–1155
2. Barker E, Barker W, Burr W, Polk W, Smid M (2007) Recommendation for key management - part 1: general. http://csrc.nist.gov/publications/nistpubs/800-57/sp800-57-Part1-revised2_Mar08-2007.pdf, March, 2007, Revised January 2016
3. Batina L, Mentens N, Sakiyama K, Preneel B, Verbauwhede I (2006) Low-cost elliptic curve cryptography for wireless sensor networks. In: Security and privacy in ad-hoc and sensor networks—ESAS 2006. Lecture notes in computer science, vol 4357. Springer, Berlin, pp 6–17
4. BBC News (2014) NSA 'developing code-cracking quantum computer'. http://www.bbc.com/news/technology-25588605
5. Bernstein DJ, Engels S, Lange T, Niederhagen R, Paar C, Schwabe P, Zimmermann R (2016) Faster elliptic-curve discrete logarithms on FPGAs. Cryptology ePrint archive, report 2016/382. http://eprint.iacr.org/2016/382
6. Bock H, Braun M, Dichtl M, Hess E, Heyszl J, Kargl W, Koroschetz H, Meyer B, Seuschek H (2008) A milestone towards RFID products offering asymmetric authentication based on elliptic curve cryptography. In: Proceedings of the 4th workshop on RFID security—RFIDSec 2008
7. Hankerson D, Menezes AJ, Vanstone S (2003) Guide to elliptic curve cryptography. Springer, New York
8. Hein D, Wolkerstorfer J, Felber N (2009) ECC is ready for RFID - a proof in silicon. In: Selected areas in cryptography—SAC 2008. Lecture notes in computer science, vol 5381. Springer, Berlin, pp 401–413
9. Koblitz N (1987) Elliptic curve cryptosystems. Math Comput 48:203–209
10. Kumar S, Paar C, Pelzl J, Pfeiffer G, Schimmler M (2006) Breaking ciphers with COPACOBANA—a cost-optimized parallel code breaker. In: Cryptographic hardware and embedded systems (CHES 2006). Lecture notes in computer science, vol 4249. Springer, pp 101–118
11. Lee YK, Sakiyama K, Batina L, Verbauwhede I (2008) Elliptic-curve-based security processor for RFID. IEEE Trans Comput 57(11):1514–1527
12. Lyubashevsky V, Peikert C, Regev O (2010) On ideal lattices and learning with errors over rings. In: Advances in cryptology - EUROCRYPT 2010. Lecture notes in computer science, vol 6110. Springer, Berlin, pp 1–23
13. National Institute of Standards and Technology (2009) Discussion paper: the transitioning of cryptographic algorithms and key sizes. http://csrc.nist.gov/groups/ST/key_mgmt/documents/Transitioning_CryptoAlgos_070209.pdf
14. National Institute of Standards and Technology (2016) NIST Kicks off effort to defend encrypted data from quantum computer threat. www.nist.gov/news-events/news/2016/04/nist-kicks-effort-defend-encrypted-data-quantum-computer-threat
15. Pessl P, Hutter M (2014) Curved tags—a low-resource ECDSA implementation tailored for RFID. In: Workshop on RFID security—RFIDSec 2014
16. Regev O (2009) On lattices, learning with errors, random linear codes, and cryptography. J ACM 56(6)
17. Rivest RL, Shamir A, Adleman L (1978) A method for obtaining digital signatures and public-key cryptosystems. Commun ACM 21(2):120–126
18. Shor PW (1994) Algorithms for quantum computation: discrete logarithms and factoring. In: Proceedings of the 35th annual symposium on foundations of computer science, SFCS '94. IEEE Computer Society, Washington, pp 124–134
19. I. T. U. Telecommunication Development Bureau (2015) ICT facts and figures. https://www.itu.int/en/ITU-D/Statistics/Documents/facts/ICTFactsFigures2015.pdf

20. The guardian (2013) The NSA files: *Decoded*. https://www.theguardian.com/us-news/the-nsa-files
21. Wenger E (2013) Hardware architectures for MSP430-based wireless sensor nodes performing elliptic curve cryptography. In: Applied cryptography and network security—ACNS 2013. Lecture notes in computer science, vol 7954. Springer, Berlin, pp 290–306
22. Wenger E, Hutter M (2011) A hardware processor supporting elliptic curve cryptography for less than 9 kGEs. In: Smart card research and advanced applications—CARDIS 2011. Lecture notes in computer science, vol 7079. Springer, Berlin, pp 182–198
23. ZDNet (2015) The internet of things and big data: unlocking the power. http://www.zdnet.com/

Chapter 2
Background

2.1 Introduction to Public-Key Cryptography

In this chapter we review the concept of public-key cryptography (PKC) and then describe two PKC schemes namely, elliptic-curve cryptography and lattice-based cryptography. PKC was introduced by Diffie and Hellman in 1976 [9]. In a PKC scheme, a user, say Bob, has a pair of keys: a widely disseminated *public-key* and a secret *private-key*. To send messages to Bob, Alice and other users use Bob's public-key. Only Bob can recover the messages from the ciphertexts by using his private-key. The basic concept of the public-key encryption is shown in Fig. 2.1.

The security of a PKC scheme is based on the assumption that it is computationally in-feasible to compute the private-key from the public-key. Such security assumption is assured by computationally hard mathematical problems such as the integer factorization problem, the discrete logarithm problem, the elliptic-curve discrete logarithm problem, and several hard problems defined over lattices etc. In this research we implement a set of PKC schemes based on the elliptic-curve discrete logarithm problem and lattice problems.

The chapter is organized as follows: in the remaining part of this section we define the elliptic-curve discrete logarithm problem and the well-known lattice problems. In Sect. 2.2 we introduce binary extension fields and Koblitz curves. The next section briefly describes the field primitives and a public-key encryption scheme that uses elliptic-curve cryptography. Cryptography schemes based on lattice problems are described in Sect. 2.4. The section also briefly describes the building blocks for implementing the lattice-based schemes. The final section gives a summary.

© Springer Nature Singapore Pte Ltd. 2020   7
S. Sinha Roy and I. Verbauwhede, *Lattice-Based Public-Key Cryptography in Hardware*,
Computer Architecture and Design Methodologies,
https://doi.org/10.1007/978-981-32-9994-8_2

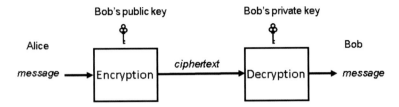

Fig. 2.1 Basic concept of public-key encryption

2.1.1   The Elliptic-Curve Discrete Logarithm Problem

In 1986 Koblitz [18] and Miller [22] independently proposed cryptography using elliptic-curves, and since then elliptic-curve cryptography (ECC) has become very popular for designing fast public-key schemes on various platforms.

Definition 2.1.1 (*Elliptic-curves* [13]) An elliptic-curve E over a field $\mathbb{K}$ is defined by a so-called Weierstrass equation

$$E : y^2 + a_1 x y + a_3 y = x^3 + a_2 x^2 + a_4 x + a_6 , \qquad (2.1)$$

where $a_1, a_2, a_3, a_4, a_6 \in \mathbb{K}$ and $\Delta \neq 0$, where Δ is the discriminant, defined as follows:

$$\Delta = -d_2^2 d_8 - 8 d_4^3 - 27 d_6^2 + 9 d_2 d_4 d_6$$
$$d_2 = a_1^2 + 4 a_2$$
$$d_4 = 2 a_4 + a_1 a_3$$
$$d_6 = a_3^2 + 4 a_6$$
$$d_8 = a_1 a_6 + 4 a_2 a_6 - a_1 a_3 a_4 + a_2 a_3^2 - a_4^2 .$$

Let $P(x, y)$ represents a point on E with coordinates (x, y). All points on the curve and a special point ∞ known as the point at infinity, form a group under the elliptic-curve addition rule in $\mathbb{K}$. Let us denote the group by $E(\mathbb{K})$. Let two points on E be $P_1(x_1, y_1)$ and $P_2(x_2, y_2)$. Their sum is the point $P_3(x_3, y_3)$ on the curve. The addition rule uses the *chord-and-tangent* method for adding two points on the curve (Fig. 2.2). The method computes additions, subtractions, multiplications, and inversions in $\mathbb{K}$ (see Sect. 2.2).

By using the point addition and doubling operations, the well-known *double-and-add* method computes a scalar multiple k of $P(x, y)$ which is again a point $Q = k \cdot P$ on E. This computation is called *scalar multiplication* or *point multiplication*. Algorithm 1 shows the double-and-add approach for point multiplication.

Here P is called the base point and Q is called the scalar multiple of the base point.

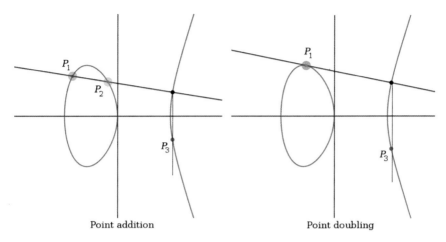

Fig. 2.2 Geometric representation of point addition and doubling using the chord-and-tangent rule

---

Algorithm 1: *Double-and-add* method for point multiplication [13]

Input: Scalar $k = \sum_{i=0}^{l-1} k_i 2^i$ and a point P on E
Output: Point $Q = k \cdot P$ on E

1 **begin**
2     $Q \leftarrow \infty$;
3   **for** $i = l - 1$ *downto* 0 **do**
4        $Q \leftarrow 2Q$;
5   **if** $k_i = 1$ **then**
6           $Q \leftarrow Q + P$;
7   **end**
8   **end**
9 **end**

Definition 2.1.2 (*The elliptic-curve discrete logarithm problem (ECDLP)*) For a given base point P on the curve E and a scalar multiple $Q = k \cdot P$, find the scalar k.

The ECDLP is considered to be a hard problem when the order of the base point has a large prime factor p. Advanced algorithms for solving ECDLP, such as Pollard's rho algorithm [25] has an expected time complexity $\mathcal{O}(\sqrt{p})$. A survey of the attack algorithms is provided by Galbraith and Gaudry [11]. The ECDLP has been used to construct key exchange schemes, public-key encryption schemes, and digital signature schemes [13].

2.1.2 Lattice Problems

In this section we review hard problems defined over lattices. Such problems are a class of optimization problems and their conjectured intractability is the foundation of lattice-based public-key cryptography schemes.

Definition 2.1.3 (*Lattice* [14]) Let $\mathcal{V}$ be a set of n linearly independent vectors $v_0, \ldots, v_{n-1} \in \mathbb{R}^m$. The lattice $\mathcal{L}$ is the set of linear combinations of the vectors with coefficients in $\mathbb{Z}$.

$$\mathcal{L} = \{a_0 \cdot v_0 + \cdots + a_{n-1} \cdot v_{n-1}\} \text{ where } a_0, \ldots a_{n-1} \in \mathbb{Z} . \tag{2.2}$$

The set $\mathcal{V}$ is a *basis* of $\mathcal{L}$, n is its rank and m is its dimension. The lattice is called a full-rank lattice if $n = m$. Indeed a basis for $\mathcal{L}$ is any set of n independent vectors that generates $\mathcal{L}$, and there are infinitely many basis for $m \geq 2$. Any two such basis are related by an integer matrix with determinant equal to ± 1.

Definition 2.1.4 (*Span* [14]) The span of the basis of a lattice $\mathcal{L}$ with basis $\mathcal{V}$ is the collection of all linear combinations $\alpha_0 \cdot v_0 + \cdots + \alpha_{n-1} \cdot v_{n-1}$, where $\alpha_0, \ldots, \alpha_{n-1} \in \mathbb{R}$.

One parameter for $\mathcal{L}$ is the length of a shortest nonzero vector. The length of a vector v in $\mathcal{L}$ is defined by its Euclidean norm $\| v \|$. The length of shortest vector is denoted by λ_1, which is the smallest radius r such that the lattice points inside a ball of radius r span a space of dimension 1.

Definition 2.1.5 (*Successive minima* [28]) Let Λ be a lattice of rank n. For $i \in \{1, \ldots, n\}$, we define the ith successive minimum as

$$\lambda_i(\Lambda) = \inf\{r \mid \dim(\mathrm{span}(\Lambda \cap \mathbf{B}(0, r))) \geq i\} ,$$

where $\mathbf{B}(0, r) = \{x \in \mathbb{R}^m \mid \| x \| \leq r\}$ is the closed ball of radius r around 0.

Definition 2.1.6 (*The shortest vector problem* (SVP [28])) Find a nonzero vector v in a lattice $\mathcal{L}$ such that $\| v \| = \lambda_1(\mathcal{L})$.

Ajtai [3] showed that the SVP with Euclidean norm is NP-hard for randomized reductions. The SVP is an α-approximation version of the SVP where one has to find a vector v_α in $\mathcal{L}$ such that $\| v_\alpha \| \leq \alpha \lambda_1(\mathcal{L})$. There is an absolute constant $\epsilon > 0$ so that the SVP is also NP-hard with $\alpha < 1 + 2^{-n^\epsilon}$ for randomized reductions [3].

Definition 2.1.7 (*The closest vector problem* (CVP) [14]) Given a vector $w \in \mathbb{R}^m$ that is not in $\mathcal{L}$, find a vector $v \in \mathcal{L}$ that is closest to w, i.e., find a vector $v \in \mathcal{L}$ that minimizes the Euclidean norm $\| w - v \|$.

Similar to the SVP, the CVP is an α-approximation version of the CVP where one has to find a vector v_α such that $\| w - v_\alpha \| \leq \alpha \| w - v \|$. The CVP is a generalization of the SVP and given an oracle for the CVP one can solve the SVP by making queries. The CVP is known to be NP-hard [21].

The SVP or CVP or their approximation versions are easy to solve if a basis comprised of orthogonal or nearly orthogonal and short vectors is known. Lattice reduction algorithms are a class of algorithms that aim to output such *good* basis from any given basis for a lattice. The LLL algorithm outputs an LLL-reduced basis in polynomial time but the with approximation factor C^n, where C is a small constant. Thus the LLL algorithm is very effective when the dimension n of the lattice is small. Algorithms that achieve close approximation (e.g. the AKS algorithm [4], the BKZ algorithm [29] etc.) run in exponential time. The inability of the lattice reduction algorithms to find a good basis in polynomial time is used as the foundation for the lattice-based cryptography schemes.

Construction of cryptographic schemes based on the hardness of lattice problems started with Ajtai's [2] seminal work where he showed average case to worst case reduction. This is of particular interest because the well-known number-theoretic problems such as the integer factorization or the ECDLP do not possess this feature. In 2005 Regev [27] introduced a new problem known as the learning with errors problem (LWE). Since its introduction, the LWE problem has become very popular for construction of a variety of schemes such as public-key encryption, key exchange, digital signature schemes and even homomorphic encryption schemes. The LWE problem is parametrized by the rank n of the lattice, an integer modulus q and an error distribution $\mathcal{X}$ over $\mathbb{Z}$. A secret vector $\mathbf{s}$ of rank n is chosen uniformly in $\mathbb{Z}_q^n$. Then samples are produced by selecting uniform random vectors $\mathbf{a}_i$ and error terms e_i from the error distribution $\mathcal{X}$ and by computing $b_i = \langle \mathbf{a}_i, \mathbf{s} \rangle + e_i \in \mathbb{Z}_q$. The LWE distribution $A_{s,\mathcal{X}}$ over $\mathbb{Z}_q^n \times \mathbb{Z}_q$ is defined as the set of tuples $(\mathbf{a}_i, b_i)$. The decision and search versions of the LWE problem are defined below.

Definition 2.1.8 (*The decision LWE problem* [27]) Distinguish with non-negligible advantage between a polynomial number of samples drawn from the LWE distribution $A_{s,\mathcal{X}}$ and the same number of samples drawn uniformly from $\mathbb{Z}_q^n \times \mathbb{Z}_q$.

Definition 2.1.9 (*The search LWE problem* [27]) Find the secret $\mathbf{s}$ given a polynomial number of samples from the LWE distribution $A_{s,\mathcal{X}}$.

Its hardness can be reduced to the hardness of the above mentioned lattice problems. In practice cryptosystems based on the LWE problem are slow as they require computations on large matrices with coefficients from $\mathbb{Z}_q$. There is a more practical variant of the LWE problem that is defined over polynomial rings and is called the *ring-LWE problem*.

The ring-LWE problem is a ring-based version of the LWE problem and was introduced by Lyubashevsky, Peikert and Regev in [20]. To achieve computational efficiency and to reduce the key size, ring-LWE uses special structured ideal lattices. Such lattices correspond to ideals in rings $R = \mathbb{Z}[\mathbf{x}]/\langle f \rangle$, where f is an irreducible polynomial of degree n. Let s be a secret uniformly random polynomial in $R_q = R/qR$. The ring-LWE distribution on $R_q \times R_q$ consists of polynomial tuples $(a_i(x), b_i(x))$, where the coefficients of a_i are chosen uniformly from $\mathbb{Z}_q$ and $b_i(x)$ is computed as a polynomial $a_i(x) \cdot s(x) + e_i(x) \in R_q$. Here e_i are error polynomials with coefficients sampled from an n-dimensional error distribution $\mathcal{X}$. The error distribution is typically a discrete Gaussian distribution. In some cases, e.g., for 2^k-power cyclotomics, this error distribution can be taken as the product of n independent discrete Gaussians, but in general $\mathcal{X}$ is more complex. One can construct

s by sampling the coefficients from $\mathcal{X}$ instead of sampling uniformly without any security implications [20].

Definition 2.1.10 (*The decision ring-LWE problem* [20]) Distinguish with non-negligible advantage between a polynomial number of samples $(a_i(x), b_i(x))$ drawn from the ring-LWE distribution and the same number of samples generated by choosing the coefficients uniformly.

Definition 2.1.11 (*The search ring-LWE problem* [20]) Find the secret polynomial $s(x)$ given a polynomial number of samples drawn from the ring-LWE distribution.

In cases where f is a cyclotomic polynomial, the difficulty [20] of the search ring-LWE problem is roughly equivalent to finding a short vector in an ideal lattice composed of polynomials from R. Note that in the case of the LWE problem, the hardness was related to solving the NP-hard SVP over *general* lattices. Though no proof exists to show equivalence between the SVP for general lattices and ideal lattices, the two cases are presumed to be equally difficult. The computational efficiency using the ring-LWE problem is obtained at the cost of the above security assumption. Cryptographic schemes based on the ring-LWE problem are fast thanks to simple polynomial arithmetic [12].

2.2 Elliptic-Curve Cryptography over $\mathbb{F}_{2^m}$

In cryptography, elliptic-curves defined over finite fields are used. The most commonly used finite fields for ECC are prime fields and binary extension fields $\mathbb{F}_{2^m}$. For hardware implementations, elliptic-curves over $\mathbb{F}_{2^m}$ are preferred since they can be implemented easily on hardware and because they achieve faster speed than prime fields. In this research we restrict ourselves to elliptic-curves over $\mathbb{F}_{2^m}$.

Definition 2.2.1 (*Elliptic-curves over binary fields*) The curve equation is

$$E : y^2 + xy = x^3 + ax^2 + b , \tag{2.3}$$

where the curve constants a and b are in $\mathbb{F}_{2^m}$.

For elliptic-curves defined over binary fields, the point addition and doubling rules are defined below.

Point addition For $P_1(x_1, y_1) \neq P_2(x_2, y_2)$, the equation for $P_3(x_3, y_3)$ is as follows.

$$x_3 = \lambda^2 + \lambda + x_1 + x_2 + a \text{ and } y_3 = \lambda(x_1 + x_3) + x_3 + y_1 . \tag{2.4}$$

Point doubling For $P_1(x_1, y_1) = P_2(x_2, y_2)$, the equation for $P_3(x_3, y_3)$ is as follows.

$$x_3 = \lambda^2 + \lambda + a \text{ and } y_3 = x_1^2 + \lambda x_3 + x_3 . \tag{2.5}$$

2.2.1 Koblitz Curves

Koblitz curves introduced by Koblitz in [17] are a special class of elliptic-curves defined by the following equation:

$$y^2 + xy = x^3 + ax^2 + 1 , \qquad a \in \{0, 1\} , \qquad (2.6)$$

with points with coordinates $x, y \in \mathbb{F}_{2^m}$. Koblitz curves offer efficient point multiplications because they allow trading computationally expensive point doublings to cheap Frobenius endomorphisms. Many standards use Koblitz curves including NIST FIPS 186-4 [24] which describes the (Elliptic-Curve) Digital Signature Standard (ECDSA) and defines five Koblitz curves NIST K-163, K-233, K-283, K-409, and K-571 over the finite fields $\mathbb{F}_{2^{163}}$, $\mathbb{F}_{2^{233}}$, $\mathbb{F}_{2^{283}}$, $\mathbb{F}_{2^{409}}$, and $\mathbb{F}_{2^{571}}$, respectively.

The Frobenius endomorphism for a point $P \in E$ is given by $\phi(P) = (x^2, y^2)$. For Koblitz curves, it holds that $\phi(P) \in E$ for all $P \in E$. Koblitz showed that $\phi^2(P) - \mu\phi(P) + 2P = \infty$ for all $P \in E$, where $\mu = (-1)^{1-a}$ [17]. Consequently, the Frobenius endomorphism can be seen as a multiplication by the complex number $\tau = (\mu + \sqrt{-7})/2$ [17].

Let the ring of polynomials in τ with integer coefficients be denoted by $\mathbb{Z}[\tau]$. For any element $u = u_{l-1}\tau^{l-1} + \cdots + u_0 \in \mathbb{Z}[\tau]$ with $u_i \in \{0, 1\}$, we can multiply any base point P by u as follows.

$$[u_{l-1}\tau^{l-1} + \cdots + u_0]P = [u_{l-1}]\tau^{l-1}P + \cdots + [u_0]P .$$

The point multiplication performs only point additions and Frobenius operations. Since the Frobenius operation is cheap, point multiplication by $u \in \mathbb{Z}[\tau]$ is faster than a point multiplication by an integer scalar of the same length. However, the ECDLP is defined for integer scalars only. Solinas showed that using the relation $-\tau^2 + \mu\tau = 2$, it is possible to map an integer scalar into an element in $\mathbb{Z}[\tau]$ with binary coefficients [30]. Such a representation is called a τ-adic expansion. Representing an integer scalar k as a τ-adic expansion $t = \sum_{i=0}^{\ell-1} t_i\tau^i$ allows computing point multiplications with a Frobenius-and-add algorithm, which is similar to the double-and-add algorithm except that point doublings are replaced by Frobenius endomorphisms.

Length reduction during integer to τ-adic conversion If a τ-adic representation is computed directly from the integer scalar k, then the length l of the τ-adic representation is approximately two times the bit-length of k. This expansion in length is a problem since it doubles the number of point additions. Solinas showed [30] that if the integer scalar is expressed as $k = \lambda(\tau^m - 1) + \rho$, then the point multiplication $k \cdot P(x, y)$ turns into $\rho \cdot P(x, y)$ as for Koblitz curves $(\tau^m - 1)P(x, y) = \infty$. The good thing is that now a τ-adic representation of ρ has length roughly equal to m [30]. The computation of ρ from k is called the scalar reduction operation.

There are several methods for performing scalar reductions. The *lazy reduction* method proposed by Brumley and Järvinen [6] uses divisions by τ.

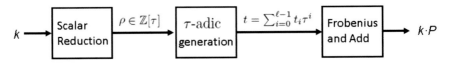

Fig. 2.3 Computation flow in point multiplication on Koblitz curves

Theorem 1 (Division by τ) *Any element* $\alpha = (d_0 + d_1\tau) \in \mathbb{Z}[\tau]$ *is divisible by* τ *if and only if* d_0 *is even. The result of the division when stored in* (d_0, d_1), *becomes*

$$(d_0, d_1) \leftarrow (\mu d_0/2 + d_1, -d_0/2) .$$

Since the division by τ can be performed by simple shift and addition operations, the scalar reduction method of Brumley and Järvinen is suitable for lightweight hardware implementations. In Chap. 3 we use this scalar reduction method for our hardware implementation. The steps to perform point multiplication on Koblitz curves are shown in Fig. 2.3.

ECIES As an example of ECC-based schemes, we describe the Elliptic-Curve Integrated Encryption Scheme (ECIES). The scheme was invented by Abdalla, Bellare and Rogaway [1] for the purpose of public-key encryption. Besides an elliptic-curve point multiplier, the ECIES scheme uses a key derivation function (KDF), a symmetric-key encryption/decryption algorithm ENC/DEC, and a message authentication code (MAC). The ECIES scheme is described as follows.

1. ECIES.KeyGen(E): For a public base point P on E, choose a random private-key k and compute the public-key $Q = k \cdot P$.
2. ECIES.Encrypt(Q, m): Randomly generate an integer r and compute the point multiplications $R_1 = r \cdot P$ and $R_2 = r \cdot Q$. Generate $(k_1, k_2) \leftarrow \text{KDF}(x_{R_2}, R_1)$ where x_{R_2} is the x-coordinate of R_2. Compute the ciphertext $c \leftarrow \text{ENC}(k_1, m)$ and compute the tag $t \leftarrow \text{MAC}(k_2, c)$. Output R_1, c and t.
3. ECIES.Decrypt(k, R_1, c, t): Compute $R_3 = k \cdot R_1$ and then use the x-coordinate of R_3 i.e., x_{R_3} as the shared secret. Note that $R_3 = k \cdot R_1 = kr \cdot P = r \cdot Q = R_2$. Hence $x_{R_3} = x_{R_2}$. Now follow the same steps as described in ECIES.Encrypt and compute (k_1, k_2) as the keys for the DEC and the MAC. Now it is trivial to recover m from c using $\text{DEC}(k_1, c)$ and then validate the recovered message.

From the above scheme we see that the elliptic-curve point multiplication plays a central role in the ECIES scheme. In this research, we restrict ourselves to the implementation of the elliptic-curve point multiplication primitive.

2.3 Primitives for Arithmetic in $\mathbb{F}_{2^m}$

Let $f(x) = x^m + \bar{f}(x)$ be the irreducible binary polynomial of degree m for the binary extension field $\mathbb{F}_{2^m}$. There are two popular ways to represent the field elements: the polynomial basis and the normal basis representations. In this research we use the polynomial basis representation. In this representation an element $a \in \mathbb{F}_{2^m}$ is represented as a polynomial $a(\theta) = \sum_{i=0}^{m-1} a_i \theta^i$ where the coefficients a_i are from $\mathbb{F}_2$ and θ is a root of $f(x)$. Addition or subtraction of two elements is coefficient-wise addition or subtraction in $\mathbb{F}_2$. In the following we review the field reduction, multiplication, squaring and inversion operations.

2.3.1 Reduction

In the polynomial basis representation when two field elements are multiplied, the result, say $c(\theta)$, is a polynomial of degree at most $2m - 2$. Since $f(x) = x^m + \bar{f}(x)$, we have $\theta^m = \bar{f}(\theta)$ in $\mathbb{F}_{2^m}$. Note that $\bar{f}(\theta)$ has a degree at most $m - 1$, and hence θ^m gets reduced to $\bar{f}(\theta)$. A naive approach to reduce $c(\theta)$ is to reduce the coefficients of θ^i sequentially for all $i \in [m, 2m - 2]$. This method is slow due to its sequential nature. A speedup is possible by reducing more than one coefficients in every iteration.

We can do a lot better if $f(x)$ is sparse. NIST has recommended $\mathbb{F}_{2^m}$ fields with irreducible polynomials having three or five nonzero coefficients [23]. For such sparse irreducible polynomials, each coefficient of the reduced result can be expressed as a boolean expression of a few input coefficients. Hence the output coefficients can be computed directly by evaluating the small boolean expressions. In this research we use the NIST recommended K-283 curve over $\mathbb{F}_{2^{283}}$ with $f(x) = x^{283} + x^{12} + x^7 + x^5 + 1$. The steps using this irreducible polynomial are shown in Algorithm 2.43 in [13].

2.3.2 Multiplication

The field multiplication in the polynomial basis representation is the multiplication of two polynomials, followed by a reduction by $f(x)$. For two elements a and $b \in \mathbb{F}_{2^m}$ the result is $\sum_{i=0}^{m-1} b(\theta) a_i \theta^i \mod f(x)$. There are several ways to compute the polynomial multiplication efficiently. A detailed description of these methods can be found in [13].

The naive method is the classical shift-and-add method for polynomial multiplication. This method has a quadratic time complexity. The time requirement can be reduced by a factor by processing the operands in a word-serial way. An efficient way to perform word-serial processing is known as the *comb method* [13].

For fast computation, the Karatsuba method is the most efficient one thanks to its $\mathcal{O}(m^{\lg_2 3})$ time complexity. The Karatsuba method splits each input operand into two polynomials of length $\lceil m/2 \rceil$. For e.g., a is split into two half-size polynomials a_h and a_l such that $a = a_h \theta^{\lceil m/2 \rceil} + a_l$. After this splitting, the multiplication is performed as

$$a \cdot b = a_h \cdot b_h \theta^{2\lceil m/2 \rceil} + \left[(a_h + a_l) \cdot (b_h + b_l) + a_h \cdot b_h + a_l \cdot b_l \right] \theta^{\lceil m/2 \rceil} + a_l \cdot b_l \ .$$

In a similar fashion, each of the small multiplications can also be split into three smaller multiplications. This gives a recursive algorithm for computing multiplication. The Karatsuba method achieves good performance for bit-parallel implementation on hardware platforms [26]. However the recursive structure of the algorithm and additional storage requirement for the partial products are costly for lightweight implementations. In this research we use the word-serial comb method [13] to perform field multiplication.

2.3.3 Squaring

A squaring in $\mathbb{F}_{2^m}$ can be computed much faster than a multiplication. In the polynomial basis representation the square of $a = \sum_{i=0}^{m-1} a_i \theta^i$ is $a^2 = \sum_{i=0}^{m-1} a_i \theta^{2i}$. The squaring operation spreads out the input coefficients by inserting zeros between each input coefficients. On hardware platforms, the spreading out of the input bits can be implemented free of cost. The only cost is due to the modular reduction by $f(x)$. Still this cost is small as we use a sparse $f(x)$.

2.3.4 Inversion

The inverse of an element $a \in \mathbb{F}_{2^m}$ is the unique element, denoted as $a^{-1} \in \mathbb{F}_{2^m}$, such that $a \cdot a^{-1} \equiv 1 \mod f(x)$. Inversion is considered to be the costliest field operation. The most commonly used methods are based on the Extended Euclidean Algorithm (EEA) or the Fermat's Little Theorem (FLT) [13].

The EEA computes the greatest common divisor (GCD) of two polynomials a and b by finding two polynomials g and h such that $a \cdot g + b \cdot h = d$ where $d = \text{GCD}(a, b)$. This property of the EEA is used to compute the inverse of an element. Since $f(x)$ is irreducible in $\mathbb{F}_{2^m}$, the GCD of a and f is always one. Hence the EEA computes g and h such that $a \cdot g + f \cdot h = 1$, and with this we get $a \cdot g \equiv 1 \mod f(x)$. Naturally $a^{-1} = g$ in $\mathbb{F}_{2^m}$.

Inversion using the FLT computes $a^{-1} = a^{2^m - 2}$. The computation requires exponentiation of the input by $2^m - 2$. Itoh and Tsujii [15] used addition chains to compute the exponentiation efficiently. The advantage of the FLT-based method over the EEA is that it requires only multiplications and squarings for the exponentiation. For hardware implementations, the FLT-based inversion method is well suited as it can reuse the multiply and square primitives. We use the FLT-based inversion for our lightweight implementation.

2.4 Ring-LWE-Based Cryptography

The ring-LWE problem has been used to construct a wide range of schemes such as public-key encryption, key exchange, digital signature and homomorphic encryption schemes. In this research we deal with ring-LWE-based public-key encryption and homomorphic schemes. We review these schemes in the remaining part of this section.

2.4.1 The LPR Public-Key Encryption Scheme

An elegant public-key encryption scheme was constructed by Lyubashevsky, Peikert, and Regev in the full version of [20] based on the ring-LWE problem. The LPR encryption scheme performs simple polynomial arithmetic such as polynomial multiplications, additions and subtractions, along with sampling from an error distribution typically a discrete Gaussian distribution χ_σ with a small standard deviation σ. It uses a global polynomial $a \in R_q$. The key generation, encryption and decryption are as follows.

1. LPR.KeyGen(a): Choose two polynomials $r_1, r_2 \in R_q$ from $\mathcal{X}_\sigma$ and compute $p = r_1 - a \cdot r_2 \in R_q$. The public-key is (a, p) and the private-key is r_2. The polynomial r_1 is simply noise and is no longer required after key generation.
2. LPR.Encrypt(a, p, m): The message m is first encoded to $\bar{m} \in R_q$. In the simplest type of encoding scheme a message bit is encoded as $(q - 1)/2$ if the message bit is 1 and 0 otherwise. Three noise polynomials $e_1, e_2, e_3 \in R_q$ are sampled from a discrete Gaussian distribution with standard deviation σ. The ciphertext then consists of two polynomials $c_1 = a \cdot e_1 + e_2$ and $c_2 = p \cdot e_1 + e_3 + \bar{m} \in R_q$.
3. LPR.Decrypt(c_1, c_2, r_2): Compute $m' = c_1 \cdot r_2 + c_2 \in R_q$ and recover the original message m from m' using a decoder. In the simplest decoding scheme the coefficients m'_i of m' are decodes as 1 if they are in the interval $(q/4, 3q/4)$, and as 0 otherwise.

A block level view of the LPR encryption and decryption is shown in Fig. 2.4.

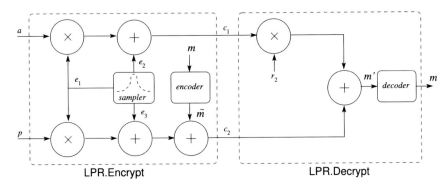

Fig. 2.4 Block level `LPR.Encrypt` and `LPR.Decrypt`

2.5 Primitives for Ring-LWE-Based Cryptography

From the descriptions of the ring-LWE-based public-key encryption and homomorphic encryption schemes, we see that the main primitives that we need are: discrete Gaussian sampling for the generation of the error polynomials, polynomial arithmetic unit for polynomial addition, subtraction and multiplication, and a division-and-round unit for computing homomorphic multiplications. These primitives are described as follows.

2.5.1 Discrete Gaussian Sampler

Definition 2.5.1 (*Discrete Gaussian distribution*) The discrete Gaussian distribution $D_{\mathbb{Z},\sigma}$ over $\mathbb{Z}$ with mean 0 and standard deviation $\sigma > 0$ is defined by $D_{\mathbb{Z},\sigma}(E = z) = \frac{1}{S} e^{-(z)^2/2\sigma^2}$, where E is a random variable on $\mathbb{Z}$ and S is the normalization factor equal to $1 + 2 \sum_{z=1}^{\infty} e^{-z^2/2\sigma^2}$ which is approximately $\sigma\sqrt{2\pi}$.

In the same way we can define a discrete Gaussian distribution $D_{\mathcal{L},\sigma}$ over a lattice $\mathcal{L}$. It assigns a probability proportional to $e^{-|\mathbf{v}|^2/2\sigma^2}$ to each element $\mathbf{v} \in \mathcal{L}$. Specifically when $\mathcal{L} = \mathbb{Z}^n$, the discrete Gaussian distribution is the product distribution of n independent copies of $D_{\mathbb{Z},\sigma}$.

Tail bound of a discrete Gaussian distribution: The tail of the Gaussian distribution is infinitely long and cannot be covered by any sampling algorithm. Indeed we need to sample up to a bound known as the *tail bound*. A finite tail-bound introduces a statistical difference with the true Gaussian distribution. The tail-bound depends on the maximum statistical distance allowed by the security parameters. As per Lemma 4.4 in [19], for any $c > 1$ the probability of sampling $\mathbf{v}$ from $D_{\mathbb{Z}^m,\sigma}$ satisfies the following inequality.

$$Pr(|\mathbf{v}| > c\sigma\sqrt{m}) < c^m e^{\frac{m}{2}(1-c^2)} . \tag{2.7}$$

Precision bound of a discrete Gaussian distribution: The probabilities in a discrete Gaussian distribution have infinitely long binary representations and hence no algorithm can sample according to a true discrete Gaussian distribution. Secure applications require sampling with high precision to maintain negligible statistical distance from the actual distribution. Let ρ_z denote the true probability of sampling $z \in \mathbb{Z}$ according to the distribution $D_{\mathbb{Z},\sigma}$. Assume that the sampler selects z with probability p_z where $|p_z - \rho_z| < \epsilon$ for some error-constant $\epsilon > 0$. Let $\tilde{D}_{\mathbb{Z},\sigma}$ denote the approximate discrete Gaussian distribution corresponding to the finite-precision probabilities p_z. The approximate distribution $\tilde{D}_{\mathbb{Z}^m,\sigma}$ corresponding to m independent samples from $\tilde{D}_{\mathbb{Z},\sigma}$ has the following statistical distance Δ to the true distribution $D_{\mathbb{Z}^m,\sigma}$ [10]:

$$\Delta(\tilde{D}_{\mathbb{Z}^m,\sigma}, D_{\mathbb{Z}^m,\sigma}) < 2^{-k} + 2mz_t\epsilon . \tag{2.8}$$

Here $Pr(|\mathbf{v}| > z_t : \mathbf{v} \leftarrow D_{\mathbb{Z}^m,\sigma}) < 2^{-k}$ represents the tail bound.

Methods for Discrete Gaussian Sampling

There are various methods for sampling from a discrete non-uniform (and also Gaussian) distribution [8]. Here we review most of these methods.

The rejection sampling is one of the simplest methods for sampling from a discrete nonuniform distribution. To sample from a target distribution T, the rejection sampling method first samples a value z from some easy proposal distribution. Then the sampled value is accepted with a probability proportional to $T_{\mathbb{Z},\sigma}(E = z)$. Though the method is simple in nature, in practice, rejection sampling for a discrete Gaussian distribution is slow due to the high rejection rate for the sampled values that are far from the center of the distribution [12]. Moreover, for each trial, many random bits are required which is very time consuming on a constrained platform.

For continuous Gaussian distributions, the Ziggurat method is very efficient to minimize the rejection rate. Buchmann, Cabarcas, Göpfert, Hülsing and Weiden proposed a discrete version of the Ziggurat method [7] to sample from discrete Gaussian distributions. Similar to the well known continuous Ziggurat method, the discrete version divides the target distribution into several rectangles distributions. The rectangles are ordered with respect to the x-coordinates of their right edges. During a sampling operation, first a rectangle is randomly chosen, and then a random coordinate x-coordinate is generated within the rectangle. The random x-coordinate is accepted as the sample output if it is also within the rectangle that is just before the randomly chosen rectangle in the ordered list. When the condition is not satisfied, a random y coordinate is generated and a costly $exp()$ computation is performed. The authors showed that discrete Ziggurat could be a good choice to reduce memory requirement when the standard deviation is large.

The inversion sampling method first generates a random probability and then selects a sample value such that the cumulative distribution up to that sample point is just larger than the randomly generated probability. To implement discrete Gaussian sampling using the inversion method, the cumulative distribution function (CDF) table CDT is precomputed with necessary tail and precision bound. The table has the property $\text{CDT}[i+1] - \text{CDT}[i] = D_{\mathbb{Z},\sigma}(E = i)$. During a sampling operation, a random number r is generated uniformly and then the CDT table is searched to find an index z such that $\text{CDT}[z] \leq r < \text{CDT}[z+1]$. The output from the sampling operation is z. Note that the bit width of the random number r should be equal to the precision bound of the distribution. As a result, this method requires a large number of random bits.

The Knuth–Yao sampling [16] uses a random walk model to generate samples from a known nonuniform discrete distribution. For the known distribution a rooted binary tree known as the discrete distribution generating (DDG) tree is constructed. The DDG tree consists of two types of nodes: intermediate nodes (I) and terminal nodes. A terminal node contains a sample point, whereas an intermediate node generates two child nodes in the next level of the DDG tree. During a sampling operation a random walk is performed starting from the root of the DDG tree. For every jump from one level of the DDG tree to the next level, a random bit is used to determine a child node. The sampling operation terminates when the random walk hits a terminal node. The value of the terminal node is the value of the output sample point. The Knuth–Yao algorithm performs sampling from non-uniform distributions using a near-optimal number of random bits.

A detailed comparative analysis of different sampling methods can be found in [10]. In this research we use the Knuth–Yao method to design a discrete Gaussian sampler for the above mentioned public-key encryption and homomorphic encryption schemes. We will review the Knuth–Yao random walk in detail in Chap. 4.

2.5.2 Polynomial Arithmetic

The ring-LWE-based schemes in Sect. 2.4 perform polynomial arithmetic in $R_q = \mathbb{Z}_q[\mathbf{x}]/\langle f \rangle$. Polynomial addition are subtraction can be performed in $\mathcal{O}(n)$ time simply by performing coefficient-wise additions or subtractions modulo q. Computation of polynomial multiplication is the costliest operation in the ring-LWE-based cryptographic schemes. In the general case, given two polynomials $a(x)$ and $b(x)$, first their product $c'(x) = a(x) \cdot b(x)$ is computed and then the product is reduced modulo $f(x)$ to get the polynomial multiplication result $c(x) = c'(x) \mod f(x)$. In the following we briefly review some of the well-known polynomial multiplication methods for computing $a(x) \cdot b(x)$. A survey of fast multiplication algorithms can be found in [5].

The school-book polynomial multiplication is the simplest method for computing the product of two polynomials. It computes the result as a summation of products using as follows.

$$a(x) \cdot b(x) = \sum_{i=0}^{n-1} \sum_{j=0}^{n-1} a_i \cdot b_j x^{i+j} . \tag{2.9}$$

From the above equation it is clear that the school-book method has $\mathcal{O}(n^2)$ time complexity. The simplicity of this method makes it attractive for designing a compact polynomial multiplier for small n. For the ring-LWE-based schemes in Sect. 2.4, n is large and hence the school-book multiplication is not suitable.

Karatsuba polynomial multiplication uses a divide-and-conquer approach to perform polynomial multiplication. For n a power of two, the Karatsuba method divides the input polynomials into polynomials of half length. E.g. $a(x)$ is split into two polynomials a_h and a_l each having $n/2$ coefficients.

$$a_h = a_{n-1}x^{n-1} + \cdots + a_{n/2}x^{n/2}$$
$$a_l = a_{n/2-1}x^{n/2-1} + \cdots + a_0 .$$

Similarly b_h and b_l are obtained after splitting $b(x)$. The multiplication is performed as follows.

$$a(x) \cdot b(x) = a_h \cdot b_h x^n + \big[(a_h + a_l) \cdot (b_h + b_l) - a_h \cdot b_h - a_l \cdot b_l\big]x^{n/2} + a_l \cdot b_l .$$

Thus the Karatsuba method has turned the n-coefficient polynomial multiplication into three $n/2$-coefficient polynomial multiplications. Following the same strategy, each of these smaller multiplications can be turned into even smaller multiplications. The Karatsuba method-based polynomial multiplication algorithms use recursive function calls to realize the divide-and-conquer strategy. The method has $\mathcal{O}(n^{\lg_2 3})$ time complexity.

The Fourier transform method is the most efficient method for computing polynomial multiplication. The n-point forward Discrete Fourier Transform (DFT) of a polynomial $a(x) = \sum_{k=0}^{n-1} a_k x^k$ consists of n evaluations of $a(x)$ at n distinct points. The good thing is that, polynomial multiplication in the Fourier domain turns into a coefficient-wise multiplication operation. Hence, if the input polynomials are provided in their Fourier representation, then we can multiply them in $\mathcal{O}(n)$ time. However the trivial way to compute the DFT has quadratic time complexity.

The Fast Fourier Transform (FFT) is an efficient way to compute the DFT of a polynomial in $O(n \lg n)$ time. It evaluates the input polynomial in the points w_n^k for the integer $k \in [0, n-1]$, where w_n is the n-th primitive root of the unity. The FFT applies a divide-and-conquer approach for the evaluation by exploiting a special property $w_n^{2k} = w_{n/2}^k$ of w_n. For n a power of two, the input polynomial $a(x)$ is split into two smaller polynomials $a_e(x) = a^{n-2}x^{n/2-1} + \cdots + a_0$ and $a_o(x) = a^{n-1}x^{n/2-1} + \cdots + a_1$ such that $a(x) = a_e(x^2) + xa_o(x^2)$. The evaluations are performed only at the $n/2$ distinct points $w_{n/2}^k$ for $k \in [0, n/2-1]$. Now the evaluation of $a(x)$ is obtained by combining the evaluations of $a_e(x)$ and $a_o(x)$ as shown below and by using the special property of ω_n i.e. $\omega_n^{2k} = \omega_{n/2}^k$ for any $k > n/2$.

$$a(\omega_n^k) = a_e(\omega_n^{2k}) + \omega_n^k a_o(\omega_n^{2k}) \ .$$

In the FFT method, the n-th primitive root of unity ω_n is a complex number, and hence FFT involves floating point arithmetic. The Number Theoretic Transform (NTT) corresponds to an FFT where the roots of unity are taken from a finite ring $\mathbb{Z}_q$. Hence all computations in NTT are performed on integers. The NTT exists if and only if n divides $d - 1$ for every prime divisor d of q. In this research we use the NTT to efficiently compute polynomial multiplication in R_q. In Chap. 5 we review an inplace iterative version of the NTT algorithm. For two polynomials a and b, their multiplication using the NTT can be computed as follows.

$$a \cdot b = INTT_\omega^{2n}\big(NTT_\omega^{2n}(a) * NTT_\omega^{2n}(b)\big) \ .$$

Here $NTT_\omega^{2n}(\cdot)$ stands for a $2n$-point NTT, $INTT_\omega^{2n}(\cdot)$ stands for the inverse transform, and the operator $*$ stands for coefficient-wise multiplications.

Reduction Modulo $f(x)$

In a polynomial multiplication over R_q, the result $c'(x) = a(x) \cdot b(x)$ which has $2n$ coefficients, is reduced modulo the irreducible polynomial $f(x)$. In the general case, i.e. when $f(x)$ does not possess any special structure, we can compute the modular reduction by computing the quotient $quo(x)$ and remainder $rem(x)$ polynomials such that $c'(x) = quo(x) \cdot f(x) + rem(x)$. The computation of $quo(x)$ requires a polynomial division operation. An efficient way to compute this division is to use the Newton iteration method. The steps are as follows [31].

Let $\mathrm{rev}_k()$ be a function that reverses the positions of the first k coefficients of the input polynomial. For e.g., when $\mathrm{rev}_3()$ is applied to the polynomial $3x^2 + 5x + 2$, the result is the polynomial $2x^2 + 5x + 3$.

From the equation $c' = quo \cdot f + rem$, where c' has degree $2n - 1$ and f has degree n, we get the relation

$$\mathrm{rev}_{2n-1}(c') = \mathrm{rev}_{n-1}(quo) \cdot \mathrm{rev}_n(f) + x^n \mathrm{rev}_{n-1}(rem) \ .$$

Therefore, we can get the congruence relation

$$\mathrm{rev}_{n-1}(quo) \equiv \mathrm{rev}_{2n-1}(c') \cdot \mathrm{rev}_n(f)^{-1} \pmod{x^n} \ .$$

For arithmetic in R_q, the irreducible polynomial f is always constant and hence $\mathrm{rev}_n(f)^{-1} \pmod{x^n}$ is a constant polynomial with n coefficients. We see that $\mathrm{rev}_{n-1}(quo) \pmod{x^n}$ can be computed by performing simple polynomial multiplication and then taking the least n coefficients of the result. Computation of quo from $\mathrm{rev}_{n-1}(quo)$ requires another application of $\mathrm{rev}_{n-1}()$. Now the remainder is computed as $rem(x) = c'(x) - quo(x) \cdot f(x)$.

2.6 Summary

In this chapter, we revisited the elliptic-curve discrete logarithm problem and hard lattice problems, and introduced a set of PKC schemes based on these problems. For elliptic-curve cryptography over binary extension fields, we described the mathematics of the finite field primitives and introduced the Koblitz curves for faster point multiplication. For lattice-based cryptography over polynomial rings, we briefly introduced the mathematics of polynomial arithmetic and discrete Gaussian sampling, and described the well-known methods for implementing these primitives.

In the next chapter we will design a lightweight Koblitz curve point multiplier and we will perform several lightweight optimizations to implement the scalar conversion. Then, from Chap. 4 onwards we will optimize the primitives for the ring-LWE-based PKC and design hardware architectures for the PKC schemes.

References

1. Abdalla M, Bellare M, Rogaway P (2001) The oracle Diffie-Hellman assumptions and an analysis of DHIES. In: Topics in cryptology–CT-RSA 2001: the cryptographers' track at RSA conference 2001, San Francisco, CA, USA, 8–12 April 2001. Proceedings. Springer, Berlin, pp 143–158
2. Ajtai M (1996) Generating hard instances of lattice problems (extended abstract). In: Proceedings of the 28th annual ACM symposium on theory of computing, STOC'96. New York, NY, USA, pp 99–108. ACM
3. Ajtai M (1998) The shortest vector problem in L2 is NP-hard for randomized reductions (extended abstract). In: Proceedings of the 30th annual ACM symposium on theory of computing, STOC'98. New York, NY, USA, pp 10–19. ACM
4. Ajtai M, Kumar R, Sivakumar D (2001) A sieve algorithm for the shortest lattice vector problem. In: Proceedings of the 33rd annual ACM symposium on theory of computing, STOC'01. New York, NY, USA, pp 601–610. ACM
5. Bernstein D (2008) Fast multiplication and its applications. Algorithmic Number Theory 44:325–384
6. Brumley BB, Järvinen KU (2010) Conversion algorithms and implementations for Koblitz curve cryptography. IEEE Trans Comput 59(1):81–92
7. Buchmann J, Cabarcas D, Göpfert F, Hülsing A, Weiden P (2014) Discrete ziggurat: a time-memory trade-off for sampling from a Gaussian distribution over the integers. In: Selected areas in cryptography—SAC 2013: 20th international conference, Burnaby, BC, Canada, August 14-16, 2013. Revised Selected Papers. Springer, Berlin, pp 402–417
8. Devroye L (1986) Non-uniform random variate generation. Springer, New York
9. Diffie W, Hellman ME (1976) New directions in cryptography. IEEE Trans Inf Theory IT-22(6):644–654
10. Dwarakanath N, Galbraith S (2014) Sampling from discrete Gaussians for lattice-based cryptography on a constrained device. Appl Algebra Eng Commun Comput 25(3):159–180
11. Galbraith SD, Gaudry P (2016) Recent progress on the elliptic curve discrete logarithm problem. Des Codes Cryptogr 78(1):51–72
12. Göttert N, Feller T, Schneider M, Buchmann J, Huss S (2012) On the design of hardware building blocks for modern lattice-based encryption schemes. In: Cryptographic hardware and embedded systems–CHES 2012. LNCS, vol 7428. Springer, Berlin, pp 512–529

13. Hankerson D, Menezes AJ, Vanstone S (2003) Guide to elliptic curve cryptography. Springer, New York
14. Hoffstein J, Pipher J, Silverman J (2008) An introduction to mathematical cryptography, 1st edn. Springer Publishing Company, New York, Incorporated
15. Itoh T, Tsujii S (1988) A fast algorithm for computing multiplicative inverses in $GF(2^m)$ using normal bases. Inf Comput 78(3):171–177
16. Knuth DE, Yao AC (1976) The complexity of non-uniform random number generation. Algorithms and complexity, pp 357–428
17. Koblitz N (1991) CM-curves with good cryptographic properties. In: Advances in cryptology–CRYPTO'91. Lecture notes in computer science, vol 576. Springer, Berlin, pp 279–287
18. Koblitz N (1987) Elliptic curve cryptosystems. Math Comput 48:203–209
19. Lyubashevsky V (2012) Lattice signatures without trapdoors. In: Proceedings of the 31st annual international conference on theory and applications of cryptographic techniques, EUROCRYPT'12. Springer, Berlin, pp 738–755
20. Lyubashevsky V, Peikert C, Regev O (2010) On ideal lattices and learning with errors over rings. In: Advances in cryptology–EUROCRYPT 2010. Lecture notes in computer science, vol 6110. Springer, Berlin, pp 1–23
21. Micciancio D (2001) The hardness of the closest vector problem with preprocessing. IEEE Trans Inf Theory 47(3):1212–1215
22. Miller V (1986) Uses of elliptic curves in cryptography. In: Advances in cryptology, Crypto'85, vol 218, pp 417–426
23. National Institute of Standard and Technology (2000) Federal Information Processing Standards Publication, FIPS 186–2. Digital Signature Standard
24. National Institute of Standards and Technology (2013) Digital signature standard (DSS). Federal Information Processing Standard, FIPS PUB 186-4
25. Pollard JM (1975) A Monte Carlo method for factorization. BIT Numer Math 15(3):331–334
26. Rebeiro C, Mukhopadhyay D (2008) Power attack resistant efficient FPGA architecture for Karatsuba multiplier. In: Proceedings of the 21st international conference on VLSI design, VLSID'08. Washington, DC, USA, pp 706–711. IEEE Computer Society
27. Regev O (2005) On lattices, learning with errors, random linear codes, and cryptography. In: Proceedings of the 37th annual ACM symposium on theory of computing, STOC'05. New York, NY, USA, pp 84–93. ACM
28. Regev O (2009) Lattices in computer science. Lecture notes of a course given in Tel Aviv University. http://www.cims.nyu.edu/~regev/teaching/lattices_fall_2009/
29. Schnorr CP, Euchner M (1994) Lattice basis reduction: improved practical algorithms and solving subset sum problems. Math Program 66(1):181–199
30. Solinas JA (2000) Efficient arithmetic on Koblitz curves. Des Codes Cryptogr 19(2–3):195–249
31. von zur Gathen J , Gerhard J (1999) Modern computer algebra. Cambridge University Press, New York

Chapter 3
Coprocessor for Koblitz Curves

CONTENT SOURCES:

Sujoy Sinha Roy, Kimmo Järvinen, Ingrid Verbauwhede Lightweight Coprocessor for Koblitz Curves: 283-Bit ECC Including Scalar Conversion with only 4300 Gates In *Cryptographic Hardware and Embedded Systems CHES* (2015).

Contribution: Main author.

Sujoy Sinha Roy, Junfeng Fan, Ingrid Verbauwhede Accelerating Scalar Conversion for Koblitz Curve Cryptoprocessors on Hardware Platforms In *IEEE Transactions on Very Large Scale Integration (VLSI) Systems* (2015).

Contribution: Main author.

3.1 Introduction

Koblitz curves [20] are a special class of elliptic-curves which enable very efficient point multiplications and, therefore, they are attractive for hardware and software implementations. However, these efficiency gains can be exploited only by representing scalars as specific τ-adic expansions. Most cryptosystems require the scalar also as an integer (see, e.g., ECDSA [25]). Therefore, cryptosystems utilizing Koblitz curves need both the integer and τ-adic representations of the scalar, which results in a need for conversions between the two domains.

In the literature a few research works exist that address the challenges of designing the scalar conversion operation in hardware. The first conversion architecture was proposed in [18] and was followed by [10, 11]. Still the converters were slow and had large area requirements. Brumley and Järvinen in [7] proposed an efficient conversion algorithm tailored for hardware platforms. Due to its sequential nature, the

© Springer Nature Singapore Pte Ltd. 2020
S. Sinha Roy and I. Verbauwhede, *Lattice-Based Public-Key Cryptography in Hardware*,
Computer Architecture and Design Methodologies,
https://doi.org/10.1007/978-981-32-9994-8_3

authors named the algorithm *lazy reduction*. The algorithm uses only integer addition, subtraction and shifting operations; no multi-precision divisions or multiplications are used. Later a speed optimized version of the lazy reduction, known as the *double lazy reduction* was proposed in [1] by Adikari, Dimitrov, and Järvinen. Still, the extra overhead introduced by these conversions has so far prevented efforts to use Koblitz curves in lightweight implementations. Lightweight applications that require elliptic-curve cryptography include, e.g., wireless sensor network nodes, RFID tags, medical implants, and smart cards. Such applications will have a central role in actualizing concepts such as the Internet of Things. However such applications have strict constraints on implementation resources such as power, energy, circuit area, memory, etc. Since the Koblitz curves are a class of computationally efficient elliptic-curves, they could potentially be the right choice for the lightweight applications. Indeed [4] showed that Koblitz curves result in a very efficient lightweight implementation if τ-adic expansions are already available. But the fact that the conversion is not included seriously limits possible applications of the implementation.

In this chapter we investigate a design methodology for implementing high security Koblitz curve cryptoprocessor on hardware platforms. We choose the NIST [24] recommended 283 bit Koblitz curve which offers around 140 bit security. We introduce several optimizations in the conversion of integer scalars. Finally, we design a lightweight coprocessor architecture of the 283-bit Koblitz curve point multiplier by using the lightweight scalar conversion architecture. We also include a set of countermeasures against timing attacks, simple power analysis (SPA), differential power analysis (DPA) and safe-error fault attacks.

The remaining part of the chapter is organized as follows: In Sect. 3.2 we optimize the Koblitz scalar conversion operation and introduce several lightweight countermeasures against side channel attacks. In Sect. 3.3 we describe the point multiplication operation. We use these optimization techniques to design a lightweight coprocessor architecture in Sect. 3.4. We provide synthesis results in 130 nm CMOS and comparisons to other works in Sect. 3.5. The final section summarizes the contributions.

3.2 Koblitz Curve Scalar Conversion

As described in Sect. 2.2.1, the scalar conversion is performed in two phases: first the integer scalar k is reduced to $\rho = b_0 + b_1\tau \equiv k \pmod{\tau^m - 1}$ and then the τ-adic representation t is generated from the reduced scalar ρ [26, 29, 32]. The overhead of these conversions is specifically important for efficient implementations. Another important aspect is resistance against side-channel attacks. Only SPA countermeasures are required because only one conversion is required per k. The scalar k is typically a nonce but even if it is used multiple times, t can be computed only once and stored.

3.2.1 Scalar Reduction

We choose the scalar reduction technique called lazy reduction (described as Algorithm 2) from [7]. The scalar k is repeatedly divided by τ for m number of times to get the following relation.

$$
\begin{aligned}
k &= (d_0 + d_1\tau)\tau^m + (b_0 + b_1\tau) \\
&= (d_0 + d_1\tau)(\tau^m - 1) + (b_0 + d_0) + (b_1 + d_1)\tau \\
&= \lambda(\tau^m - 1) + \rho \, .
\end{aligned}
$$

As shown in Theorem 1 in Sect. 2.2.1, this division can be implemented with shifts, additions, and subtractions. This makes the scalar reduction algorithm attractive for lightweight implementations. The τ-adic representation generated from ρ has a length at most $m + 4$ in F_{2^m} [7].

To meet the constraints of a lightweight platform, we implement the lazy reduction algorithm [7] in a word-serial fashion. Though this design decision reduces the area requirement, it increases the cycle count. Hence we optimize further the computational steps of Algorithm 2 to reduce the number of cycles. Further, we investigate side-channel vulnerability of the algorithm and propose lightweight countermeasures against SPA.

Computational Optimization

In lines 6 and 7 of Algorithm 2, computations of d_1 and a_0 require subtractions from zero. In a word-serial architecture with only one adder/subtracter circuit, they consume nearly 33% of the cycles of the scalar reduction. We use the iterative property of Algorithm 2 and eliminate these two subtractions by replacing lines 6 and 7 with the following ones:

Algorithm 2: Scalar reduction algorithm from [7]

Input: Integer scalar k
Output: Reduced scalar $\rho = b_0 + b_1\tau \equiv k \pmod{\tau^m - 1}$
1 $(a_0, a_1) \leftarrow (1, 0)$, $(b_0, b_1) \leftarrow (0, 0)$, $(d_0, d_1) \leftarrow (k, 0)$;
2 **for** $i = 0$ **to** $m - 1$ **do**
3 $\quad$ $u \leftarrow d_0[0]$;   /* lsb of d_0 is the remainder before division */
4 $\quad$ $d_0 \leftarrow d_0 - u$;
5 $\quad$ $(b_0, b_1) \leftarrow (b_0 + u \cdot a_0, b_1 + u \cdot a_1)$;
6 $\quad$ $(d_0, d_1) \leftarrow (d_1 - d_0/2, -d_0/2)$;   /* Division of (d_0, d_1) by τ */
7 $\quad$ $(a_0, a_1) \leftarrow (-2a_1, a_0 - a_1)$;
8 **end**
9 $\rho = (b_0, b_1) \leftarrow (b_0 + d_0, b_1 + d_1)$;

$$(d_0, d_1) \leftarrow (d_0/2 - d_1, d_0/2)$$
$$(a_0, a_1) \leftarrow (2a_1, a_1 - a_0) \tag{3.1}$$

However with this modification, (a_0, a_1) and (d_0, d_1) have a wrong sign after every odd number of iterations of the for-loop in Algorithm 2. It may appear that this wrong sign could affect correctness of (b_0, b_1) in line 5. Since the remainder u (in line 3) is generated from d_0 instead of the correct value $-d_0$, a wrong sign is also assigned to u. Hence, the multiplications $u \cdot a_0$ and $u \cdot a_1$ in line 5 are always correct, and the computation of (b_0, b_1) remains unaffected of the wrong signs.

After completion of the for-loop, the sign of (d_0, d_1) is wrong as m is an odd integer for secure fields. Hence, the correct value of the reduced scalar should be computed as $\rho \leftarrow (b_0 - d_0, b_1 - d_1)$.

Protection Against SPA

In line 5 of Algorithm 2, computation of new (b_0, b_1) depends on the remainder bit (u) generated from d_0 which is initialized to k. Multi-precision additions are performed when $u = 1$; whereas no addition is required when u is zero. A side-channel attacker can detect this conditional computation and can use, e.g., the techniques from [7] to reconstruct the secret key from the remainder bits that are generated during the scalar reduction.

One way to protect the scalar reduction from SPA is to perform dummy additions $(b_0', b_1') \leftarrow (b_0 + a_0, b_1 + a_1)$ whenever $u = 0$. However, such countermeasures based on dummy operations require more memory and are vulnerable to C safe-error fault attacks [12]. We propose a countermeasure inspired by the zero-free τ-adic representations from [26, 32]. A zero-free representation is obtained by generating the remainders u from $d = d_0 + d_1\tau$ using a map $\Psi(d) \rightarrow u \in \{1, -1\}$ such that $d - u$ is divisible by τ, but additionally not divisible by τ^2 (see Sect. 3.2.2). We observe that during the scalar reduction (which is basically a division by τ), we can generate the remainder bits u as either 1 or -1 throughout the entire for-loop in Algorithm 2. Because $u \neq 0$, a new (b_0, b_1) is always computed in the for-loop and protection against SPA is achieved without dummy operations. The following equation generates u by observing the second lsb of d_0 and lsb of d_1.

$$
\begin{aligned}
&\text{Case 1: If } d_0[1] = 0 \text{ and } d_1[0] = 0, \text{ then } u \leftarrow -1 \\
&\text{Case 2: If } d_0[1] = 1 \text{ and } d_1[0] = 0, \text{ then } u \leftarrow 1 \\
&\text{Case 3: If } d_0[1] = 0 \text{ and } d_1[0] = 1, \text{ then } u \leftarrow 1 \\
&\text{Case 4: If } d_0[1] = 1 \text{ and } d_1[0] = 1, \text{ then } u \leftarrow -1 \,.
\end{aligned}
\tag{3.2}
$$

The above equation takes an odd d_0 and computes u such that the new d_0 after division of $d - u$ by τ is also an odd integer.

Algorithm 3 shows our computationally efficient SPA-resistant scalar reduction algorithm. All operations are performed in a word-serial fashion. Since the remainder

---

Algorithm 3: SPA-resistant scalar reduction

Input: Integer scalar k
Output: Reduced scalar $\rho = b_0 + b_1\tau \equiv k \pmod{\tau^m - 1}$
1 $(a_0, a_1) \leftarrow (1, 0)$, $(b_0, b_1) \leftarrow (0, 0)$, $(d_0, d_1) \leftarrow (k, 0)$;
2 **if** $d_0[0] = 0$ **then**
3 $\quad\big|\quad e \leftarrow 1$;   /* Set to 1 when d_0 is even */
4 $\quad\big|\quad d_0[0] \leftarrow 1$;
5 **end**
6 **for** $i = 0$ **to** $m - 1$ **do**
7 $\quad\big|\quad u \leftarrow \Psi(d_0 + d_1\tau)$;   /* Remainder $u \in \{1, -1\}$, computed using (3.2) */
8 $\quad\big|\quad d_0 \leftarrow d_0 - u$;
9 $\quad\big|\quad (b_0, b_1) \leftarrow (b_0 + u \cdot a_0, b_1 + u \cdot a_1)$;
10 $\quad\big|\quad (d_0, d_1) \leftarrow (d_0/2 - d_1, d_0/2)$;   /* Saves one subtraction */
11 $\quad\big|\quad (a_0, a_1) \leftarrow (2a_1, a_1 - a_0)$;   /* Saves one subtraction */
12 **end**
13 $\rho = (b_0, b_1) \leftarrow (b_0 - d_0 - e, b_1 - d_1)$;   /* Subtraction */

---

generation in (3.2) requires the input d_0 to be an odd integer, the lsb of d_0 is always set to 1 (in line 4) when the input scalar k is an even integer. In this case, the algorithm computes the reduced scalar of $k + 1$ instead of k and after the completion of the reduction, the reduced scalar should be decremented by one. Algorithm 3 uses a one-bit register e to implement this requirement. The final subtraction in line 13 uses e as a borrow to the adder/subtracter circuit. In the next section, we show that the subtraction $d_0 - u$ in line 8 also leaks information about u and propose a countermeasure that prevents this.

3.2.2 Computation of τ-Adic Representation

For side-channel attack resistant point multiplication, we use the zero-free τ-adic representation proposed in [26, 32] and described in Algorithm 4. We add the following improvements to the algorithm.

Computational Optimization

Computation of b_1 in line 5 of Algorithm 4 requires subtraction from zero. Similar to Sect. 3.2.1 this subtraction can be avoided by computing $(b_0, b_1) \leftarrow (b_0/2 - b_1, b_0/2)$. With this modification, the sign of (b_0, b_1) will be wrong after an odd number of iterations. In order to correct this, the sign of t_i should be flipped for odd i (by multiplying it with $(-1)^i$).

Algorithm 4: Computation of zero-free τ-adic representation [26]

Input: Reduced scalar $\rho = b_0 + b_1\tau$ with b_0 odd
Output: Zero-free τ-adic bits $(t_{\ell-1}, \cdots t_0)$
1 $i \leftarrow 0$;
2 **while** $|b_0| \neq 1$ **or** $b_1 \neq 0$ **do**
3 $\quad$ $u \leftarrow \Psi(b_0 + b_1\tau)$;   /* Computed using (3.2) */
4 $\quad$ $b_0 \leftarrow b_0 - u$;
5 $\quad$ $(b_0, b_1) \leftarrow (b_1 - b_0/2, -b_0/2)$;
6 $\quad$ $t_i \leftarrow u$;
7 $\quad$ $i \leftarrow i + 1$;
8 **end**
9 $t_i \leftarrow b_0$;

Protection Against SPA

Though point multiplications with zero-free representations are resistant against SPA [26], the generation of τ-adic bits (Algorithm 4) is vulnerable to SPA. In line 3 of Algorithm 4, a remainder u is computed as per the four different cases described in (3.2) and then subtracted from b_0 in line 4. We use the following observations to detect the side-channel vulnerability in this subtraction and to propose a countermeasure against SPA.

1. For Case 1, 2, and 3 in (3.2), the subtractions of u are equivalent to flipping two (or one) least significant bits of b_0. Hence, actual subtractions are not computed in these cases.
2. For Case 4, subtraction of u from b_0 (i.e. computation of $b_0 + 1$) involves carry propagation. Hence, an actual multi-precision subtraction is computed in this case.
3. If any iteration of the while-loop in Algorithm 4 meets Case 4, then the new value of b_1 will be even. Hence, the while-loop will meet either Case 1 or Case 2 in the next iteration.

Based on the differences in computation, a side-channel attacker using SPA can distinguish Case 4 from the other three cases. Hence, the attacker can reveal around 25% of the bits of a zero-free representation. Moreover, the attacker knows that the following τ-adic bits are biased towards 1 instead of -1 with a probability of $1/3$.

We propose a very low-cost countermeasure that skips this special addition $b_0 + 1$ for Case 4 by merging it with the computation of the new (b_0, b_1) in Algorithm 4. In line 5, we compute a new b_0 as:

$$b_0 \leftarrow \left(\frac{b_0 + 1}{2} - b_1\right) = \left(\frac{b_0 - 1}{2} - \{b_1', 0\}\right) . \tag{3.3}$$

Since b_1 is an odd number for Case 4, we can represent it as $\{b_1', 1\}$ and subtract the least significant bit 1 from $(b_0 + 1)/2$ to get $(b_0 - 1)/2$. Since b_0 is always odd, the computation of $(b_0 - 1)/2$ is just a left-shift of b_0.

The computation of $b_1 \leftarrow (b_0 + 1)/2$ in line 5 of Algorithm 4 involves a carry propagation and thus an actual addition becomes necessary. We solve this problem by computing $b_1 \leftarrow (b_0 - 1)/2$ instead of the correct value $b_1 \leftarrow (b_0 + 1)/2$ and remembering the difference (i.e., 1) in a flag register h. Correctness of the τ-adic representation can be maintained by considering this difference in the future computations that use this wrong value of b_1. Now as per observation 3, the next iteration of the while-loop meets either Case 1 or 2. We adjust the previous difference by computing the new b_0 as follows:

$$b_0 \leftarrow \left(\frac{b_0}{2} - (b_1 + h) \right) = \left(\frac{b_0}{2} - b_1 - 1 \right) . \tag{3.4}$$

In a hardware architecture, this equation can be computed by setting the borrow input of the adder/subtracter circuit to 1 during the subtraction.

In (3.5), we show our new map $\Psi'(\cdot)$ that computes a remainder u and a new value h' of the difference flag following the above procedure. We consider $b_1[0] \oplus h$ (instead of $b_1[0]$ as in (3.2)) because a wrong b_1 is computed in Case 4 and the difference is kept in h.

Case 1: If $b_0[1] = 0$ and $b_1[0] \oplus h = 0$, then $u \leftarrow -1$ and $h' \leftarrow 0$
Case 2: If $b_0[1] = 1$ and $b_1[0] \oplus h = 0$, then $u \leftarrow 1$ and $h' \leftarrow 0$
Case 3: If $b_0[1] = 0$ and $b_1[0] \oplus h = 1$, then $u \leftarrow 1$ and $h' \leftarrow 0$
Case 4: If $b_0[1] = 1$ and $b_1[0] \oplus h = 1$, then $u \leftarrow -1$ and $h' \leftarrow 1$. $\tag{3.5}$

The same technique is also applied to protect the subtraction $d_0 - u$ in the scalar reduction in Algorithm 3.

Protection Against Timing Attack

The terminal condition of the while-loop in Algorithm 4 is dependent on the input scalar. Thus by observing the timing of the computation, an attacker is able to know the higher order bits of a short τ-adic representation. This allows the attacker to narrow down the search domain. We observe that we can continue the generation of zero-free τ-adic bits even when the terminal condition in Algorithm 4 is reached. In this case, the redundant part of the τ-adic representation is equivalent to the value of b_0 when the terminal condition was reached for the first time; hence the result of the point multiplication remains correct. For example, starting from $(b_0, b_1) = (1, 0)$, the algorithm generates an intermediate zero-free representation $-\tau - 1$ and again reaches the terminal condition $(b_0, b_1) = (-1, 0)$. The redundant representation $-\tau^2 - \tau - 1$ is equivalent to 1. If we continue, then the next terminal condition is again reached after generating another two bits. In this chapter we generate zero-free τ-adic representations that have lengths always larger than or equal to m of the field $\mathbb{F}_{2^m}$. To implement this feature, we added the terminal condition $i < m$ to the while-loop.

Algorithm 5: SPA-resistant generation of a zero-free τ-adic representation

---
Input: Reduced scalar $\rho = b_0 + b_1\tau$
Output: τ-adic bits $(t_{ell-1}, \cdots t_0)$ and flag f
1 $f \leftarrow$ assign_flag$(b_0[0], b_1[0])$;
2 $(b_0[0], b_1[0]) \leftarrow$ bitflip$(b_0[0], b_1[0], f)$;   /* Initial adjustment */
3 $i \leftarrow 0$;
4 $h \leftarrow 0$;
5 **while** $i < m$ **or** $|b_0| \neq 1$ **or** $(b_1 \neq 0$ **and** $h = 0)$ **or** $(b_1 \neq -1$ **and** $h = 1)$ **do**
6 $\quad$ $(u, h') \leftarrow \Psi'(b_0 + b_1\tau)$;   /* Computed using (3.5) */
7 $\quad$ $b_0[1] \leftarrow \neg b_0[1]$; /* Second LSB is set to 1 when Case 1 occurs */
8 $\quad$ $(b_0, b_1) \leftarrow (\frac{b_0}{2} - b_1 - h, \frac{b_0}{2})$;
9 $\quad$ $t_i \leftarrow (-1)^i \cdot u$;
10 $\quad$ $h \leftarrow h'$;
11 $\quad$ $i \leftarrow i + 1$;
12 **end**
13 $t_i \leftarrow (-1)^i \cdot b_0$;

In Algorithm 5, we describe an algorithm for generating zero-free representations that applies the proposed computational optimizations and countermeasures against SPA and timing attacks. The while-loops of both Algorithms 4 and 5 require b_0 to be an odd integer. When the input ρ has an even b_0, then an adjustment is made by adding one to b_0 and adding (subtracting) one to (from) b_1 when b_1 is even (odd). This adjustment is recorded in a flag f in the following way: if b_0 is odd, then $f = 0$; otherwise $f = 1$ or $f = 2$ depending on whether b_1 is even or odd, respectively. In the end of a point multiplication, this flag is checked and $(\tau + 1)P$ or $(-\tau + 1)P$ is subtracted from the point multiplication result if $f = 1$ or $f = 2$, respectively. This compensates the initial addition of $(\tau + 1)$ or $(-\tau + 1)$ to the reduced scalar ρ described in line 2 of Algorithm 5.

We also designed a high-speed scalar conversion architecture based on [1]. The optimization strategy and the hardware architecture is described in the Appendix A.

3.3 Point Multiplication

We base the point multiplication algorithm on the use of the zero-free representation discussed in Sect. 3.2. We give our modification of the point multiplication algorithm of [26, 32] with window size $w = 2$ in Algorithm 6. The algorithm includes countermeasures against SPA, DPA, and timing attacks as well as inherent resistance against C safe-error fault attacks. Implementation details of each operation used by Algorithm 6 are given in Appendix B. Below, we give a high-level description.

Line 1 computes the zero-free representation t given an integer k using Algorithms 3 and 5. It outputs a zero-free expansion of length ℓ with $t_i \in \{-1, +1\}$ represented as an ℓ-bit vector and a flag f. Lines 2 and 3 perform the precomputations by computing $P_{+1} = \phi(P) + P$ and $P_{-1} = \phi(P) - P$. Lines 4 and 5 initialize

---

Algorithm 6: Zero-free point multiplication with side-channel countermeasures

Input: An integer k, the base point $P = (x, y)$, a random element $r \in \mathbb{F}_{2^m}$
Output: The result point $Q = kP$

1 $(t, f) \leftarrow \text{Convert}(k)$;   /* Algorithms 3 and 5 */
2 $P_{+1} \leftarrow \phi(P) + P$;
3 $P_{-1} \leftarrow \phi(P) - P$;
4 **if** ℓ *is odd* **then** $Q = (X, Y) \leftarrow t_{\ell-1} P$; $i \leftarrow \ell - 3$;
5 **else** $Q = (X, Y) \leftarrow t_{\ell-1} P_{t_{\ell-2} t_{\ell-1}}$; $i \leftarrow \ell - 4$;
6 $Q = (X, Y, Z) \leftarrow (Xr, Yr^2, r)$;
7 **while** $i \geq 0$ **do**
8      $Q \leftarrow \phi^2(Q)$;
9      $Q \leftarrow Q + t_{i+1} P_{t_i t_{i+1}}$;
10      $i \leftarrow i - 2$;
11 **end**
12 **if** $f = 1$ **then** $Q \leftarrow Q + P_{-1}$;
13 **else if** $f = 2$ **then** $Q \leftarrow Q - P_{+1}$;
14 $Q = (X, Y) \leftarrow (X/Z, Y/Z^2)$;
15 **return** Q ;

---

the accumulator point Q depending on the length of the zero-free expansion. If the length is odd, then Q is set to $\pm P$ depending on the msb $t_{\ell-1}$. If the length is even, then Q is initialized with $\pm\phi(P) \pm P$ by using the precomputed points depending on the values of the two msb's $t_{\ell-1}$ and $t_{\ell-2}$. Line 6 randomizes Q by using a random element $r \in \mathbb{F}_{2^m}$ as suggested by Coron [8]. This randomization offers protection against DPA and attacks that calculate hypotheses about the values of Q based on its known initial value (e.g., the doubling attack [13]).

Lines 7 to 10 iterate the main loop of the algorithm by observing two bits of the zero-free expansion on each iteration. Each iteration begins in line 8 by computing two Frobenius endomorphisms. Line 9 either adds or subtracts $P_{+1} = (x_{+1}, y_{+1})$ or $P_{-1} = (x_{-1}, y_{-1})$ to or from Q depending on the values of t_i and t_{i+1} processed by the iteration. It is implemented by using the equations from [2] which compute a point addition in mixed affine and López-Dahab [23] coordinates. Point addition and subtraction are carried out with the exactly same pattern of operations (see Appendix B). Lines 11 and 12 correct the adjustments that ensure that b_0 is odd before starting the generation of the zero-free representation (see Sect. 3.2.2). Line 13 retrieves the affine point of the result point Q.

The pattern of operations in Algorithm 6 is almost constant. The side-channel properties of the conversion (line 1) were discussed in Sect. 3.2. The precomputation (lines 2 and 3) is fixed and operates only on the base point, which is typically public. The initialization of Q (lines 4 and 5) can be carried out with a constant pattern of operations with the help of dummy operations. The randomization of Q protects from differential power analysis (DPA) and comparative side-channel attacks (e.g., the doubling attack [13]). The main loop operates with a fixed pattern of operations on a randomized Q offering protecting against SPA and DPA. Lines 11 and 12 depend on t (and, thus, k) but they leak at most one bit to an adversary who can determine whether they were computed or not. This leakage can be prevented with a dummy operation. Although the algorithm includes dummy operations, it offers good protection also against C safe-error fault attacks. The reason is that the main loop does not involve any dummy operations and, hence, even an attacker, who is able to distinguish dummy operations, learns only few bits of information (at most, the lsb and the msb and whether the length is odd or even). Hence, C safe-error fault attacks that aim to reveal secret information by distinguishing dummy operations are not a viable attack strategy [12].

3.4 Architecture

In this section, we describe the hardware architecture (Fig. 3.1) of our ECC coprocessor for 16-bit microcontrollers such as TI MSP430F241x or MSP430F261x [31]. Such families of low-power microcontrollers have at least 4KB of RAM and can run at 16 MHz clock. We connect our coprocessor to the microcontroller using a

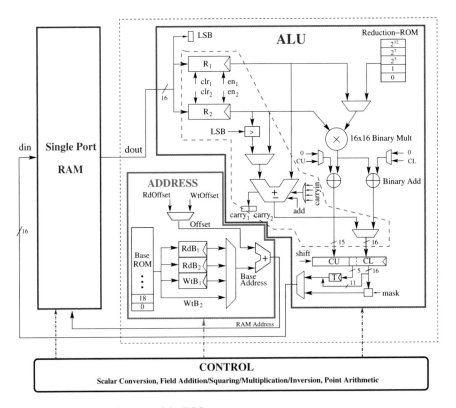

Fig. 3.1 Hardware architecture of the ECC coprocessor

memory-mapped interface [28] following the drop-in concept from [33] where the coprocessor is placed on the bus between the microcontroller and the RAM and memory access is controlled with multiplexers. The coprocessor consists of the following components: an arithmetic and logic unit (ALU), an address generation unit, a shared memory and a control unit composed of hierarchical finite state machines (FSMs).

The Arithmetic and Logic Unit (ECC-ALU)

Has a 16-bit data path and is used for both integer and binary field computations. The ECC-ALU is interfaced with the memory block using an input register pair (R_1, R_2) and an output multiplexer. The central part of the ECC-ALU consists of a 16-bit integer adder/subtracter circuit, a 16-bit binary multiplier and two binary adders. A small *Reduction-ROM* contains several constants that are used during modular reductions and multiplications by constants. The accumulator register pair (CU, CL) stores the intermediate or final results of any arithmetic operation. Finally, the output multiplexer is used to store the contents of the registers CL, T and a masked version of CL in the memory block, which sets the msb's of the most significant word of an alement to zero.

The Memory Block

Is a single-port RAM which is shared by the ECC coprocessor and the 16-bit microcontroller. Each 283-bit element of $\mathbb{F}_{2^{283}}$ requires 18 16-bit words totaling 288 bits. The coprocessor requires storage for 14 elements of $\mathbb{F}_{2^{283}}$ (see Appendix B), which gives 4032 bits of RAM (252 16-bit words). Some of these variables are reused for different purposes during the conversion.

The Address Unit

Generates address signals for the memory block. A small *Base-ROM* is used to keep the base addresses for storing different field elements in the memory. During any integer operation or binary field operation, the two address registers RdB_1 and RdB_2 in the address unit are loaded with the base addresses of the input operands. Similarly the base addresses for writing intermediate or final results in the memory block are provided in the register WtB_1 and in the output from the *Base-ROM* (WtB_2). The adder circuit of the address block is an 8-bit adder which computes the physical address from a read/write offset value and a base address.

The Control Unit

consists of a set of hierarchical FSMs that generate control signals for the blocks described above. The FSMs are described below.

(1) Scalar Conversion uses the part of the ECC-ALU shown by the red dashed polygon in Fig. 3.1. The computations controlled by this FSM are mainly integer additions, subtractions and shifts. During any addition or subtraction, the words of the operands are first loaded in the register pair (R_1, R_2). The result-word is computed using the integer adder/subtracter circuit and stored in the accumulator register CL. During a right-shift, R_2 is loaded with the operand-word and R_1 is cleared. Then the lsb of the next higher word of the operand is stored in the one-bit register LSB. Now the integer adder is used to add the shifted value $\{LSB, R_2/2\}$ with R_1 to get the shifted word. One scalar conversion requires around 78,000 cycles.

(2) Binary Field Primitives use the registers and the portion of the ECC-ALU outside the red-dashed polygon in Fig. 3.1.

- Field addition sequentially loads two words of the operands in R_2, then multiplies the words by 1 (from the *Reduction-ROM*) and finally calculates the result-word in CL after accumulation. One field addition requires 60 cycles.
- Field multiplication uses word-serial comb method [14]. It loads the words of the operands in R_1 and R_2, then multiplies the words and finally accumulates. After the completion of the comb multiplication, a modular reduction is performed requiring mainly left-shifts and additions. The left-shifts are performed by multiplying the words with the values from the *Reduction-ROM*. One field multiplication requires 829 cycles.
- Field squaring computes the square of an element of $\mathbb{F}_{2^{283}}$ in linear time by squaring its words. The FSM first loads a word in both R_1 and R_2 and then squares the word by using the binary multiplier. After squaring the words, the FSM performs a modular reduction. The modular reduction is shared with the field multiplication FSM. One field squaring requires 200 cycles.
- Field inversion uses the Itoh–Tsujii algorithm [17] and performs field multiplications and squarings following an addition chain (1, 2, 4, 8, 16, 17, 34, 35, 70, 140, 141, 282) for $\mathbb{F}_{2^{283}}$. One inversion requires 65,241 cycles.

(3) Point Operations and Point Multiplication are implemented by combining an FSM with a hardwired program ROM. The program ROM includes subprograms for all operations of Algorithm 6 and the address of the ROM is controlled by the FSM in order to execute Algorithm 6 (see Appendix B for details).

Algorithm 6 is executed so that the microcontroller initializes the addresses reserved for the accumulator point Q with the base point (x, y) and the random element r by writing $(X, Y, Z) \leftarrow (x, y, r)$. The scalar k is written into the RAM before the microcontroller issues a start point multiplication command. When this command is received, the reduction part of the conversion is executed followed by the computation of the msb(s) of the zero-free expansion. After this, the precomputations are performed by using (x, y) and the results are stored into the RAM. The initialization of Q is performed by writing either P_{+1} or P_{-1} in (X, Y) if the length of the expansion is even; otherwise, a dummy write is performed. Similarly, the sign of Q is changed if $t_{\ell-1} = -1$ and a dummy operation is computed otherwise. The main loop first executes two Frobenius endomorphisms and, then, issues an instruction that computes the next two bits of the zero-free expansion. By using these bits, either a point addition or a point subtraction is computed with P_{+1} or P_{-1}. One iteration of the main loop takes 9537 clock cycles. In the end, the affine coordinates of the result point are retrieved and they become available for the microcontroller in the addresses for the X and Y coordinates of Q.

3.5 Results and Comparisons

We described the architecture of Sect. 3.4 by using mixed Verilog and VHDL and simulated it with ModelSim SE 6.6d. We synthesized the code with Synopsys Design Compiler D-2010.03-SP4 using the regular compile for UMC 130 nm CMOS with voltage of 1.2 V by using Faraday FSC0L low-leakage standard cell libraries. The area given by the synthesis is 4,323 GE including everything in Fig. 3.1 except the single-port RAM. Computing one point multiplication requires in total 1,566,000 clock cycles including the scalar conversion. The power consumption at 16 MHz is 97.70 µW which gives an energy consumption of approximately 9.56 µJ per point multiplication. Table 3.1 summarizes our synthesis results together with several other lightweight ECC implementations from the literature. Since several of the reported implementations do not provide details of the libraries they used, we mention only the CMOS technology in the table. For a similar technology, there will be small variations in the area, power and energy consumption for different libraries.

Among all lightweight ECC processors available in the literature, the processor from [4] is the closest counterpart to our implementation because it is so far the only one that uses Koblitz curves. Even it has many differences with our architecture which make fair comparison difficult. The most obvious difference is that the processor from [4] is designed for a less secure Koblitz curve NIST K-163. Also the architecture of [4] differs from ours in many fundamental ways: they use a finite field over normal basis instead of polynomial basis, they use a bit-serial multiplier that requires all bits of both operands to be present during the entire multiplication instead of a word-serial architecture that we use, they store all variables in registers embedded into the processor architecture instead of an external RAM, and they also do not provide support for scalar conversions or any countermeasures against side-channel attacks. They also provide implementation results on 65 nm CMOS. Our architecture is significantly more scalable for different Koblitz curves because, besides control logic and RAM requirements, other parts remain almost the same, whereas the entire multiplier needs to be changed for [4]. It is also hard to see how scalar conversions or side-channel countermeasures could be integrated into the architecture of [4] without significant increases on both area and latency.

Table 3.1 Comparison to other lightweight coprocessors for ECC. The top part consists of relevant implementations from the literature. We also provide estimates for other parameter sets in order to ease comparisons to existing works

Work	Curve	Conv.	RAM	Tech. (nm)	Freq. (MHz)	Area (GE)	Latency (cycles)	Latency (ms)	Power (μW[a])	Energy (μJ[a])
[5], 2006	B-163	n/a	No	130	0.500	9,926	95,159	190.32	<60	<5.7
[6], 2008	B-163	n/a	Yes	220	0.847	12,876	–	95	93	7.48
[15], 2008	B-163	n/a	Yes	180	0.106	13,250	296,299	2,792	80.85	23.9
[21], 2006	B-163	n/a	Yes	350	13.560	16,207	376,864	27.90	n/a	n/a
[22], 2008	B-163	n/a	Yes	130	1.130	12,506	275,816	244.08	32.42	8.94
[34], 2011	B-163	n/a	Yes	130	0.100	8,958	286,000	2,860	32.34	9.25
[33], 2013	B-163	n/a	No	130	1.000	4,114	467,370	467.37	66.1	30.9
[27], 2014	P-160	n/a	Yes	130	1.000	12,448[b]	139,930	139.93	42.42	5.93
[4], 2014	K-163	no	Yes[c]	65	13.560	11,571	106,700	7.87	5.7	0.6
Our, est.	B-163	Yes	No	130, Faraday	16.000	≈3,773	≈485,000	≈30.31	≈6.11	2.96
Our, est.	K-163	Yes	No	130, Faraday	16.000	≈4,323	≈420,900	≈26.30	≈6.11	2.57
Our, est.	B-283	Yes	No	130, Faraday	16.000	≈3,773	≈1,934,000	≈120.89	≈6.11	11.8
Our, est.	K-283	Yes	Yes[d]	130, Faraday	16.000	10,204	1,566,000	97.89	>6.11	>9.6
Our	**K-283**	**yes**	**no**	**130**	**16.000**	**4,323**	**1,566,000**	**97.89**	**6.11**	**9.6**

[a]Normalized to 1 MHz

[b]Contains everything required for ECDSA including a Keccak module

[c]All variables are stored in registers inside the processor

[d]The 256 × 16-bit RAM is estimated to have an area of 5794 GE because the size of a single-port 256 × 8-bit RAM has an area of 2897 GE [33]

Note For a similar technology, there can be small variations in area, power, and energy depending on the library

Table 3.1 includes also implementations that use the binary curve B-163 and the prime curve P-160 from [25]. The area of our coprocessor is on the level of the smallest coprocessors available in the literature. Hence, the effect of selecting a 283-bit elliptic-curve instead of a less secure curve is negligible in terms of area. The price to pay for higher security comes in the form of memory requirements and computation latency. The amount of memory is not a major issue because our processor shares the memory with the microcontroller which typically has a large memory (e.g. TI MSP430F241x and MSP430F261x have at least 4KB RAM [31]). Also the computation time is on the same level with other published implementations because our coprocessor is designed to run on the relatively high clock frequency of the microcontroller which is 16 MHz.

In this work our main focus was to investigate feasibility of lightweight implementations of Koblitz curves for applications demanding high security. To enable a somewhat fair comparison with the existing lightweight implementations over $\mathbb{F}_{2^{163}}$, Table 3.1 provides estimates for area and cycles of ECC coprocessors that follow the design decisions presented in this chapter and perform point multiplications on curves B-163 or K-163. Our estimated cycle count for scalar multiplication over $\mathbb{F}_{2^{163}}$ is based on the following facts:

1. A field element in $\mathbb{F}_{2^{163}}$ requires 11 16-bit words, and hence, is smaller by a factor of 0.61 than a field element in $\mathbb{F}_{2^{283}}$. Since field addition and squaring have linear complexity, we estimate that the cycle counts for these operations scale down by a factor of around 0.61 and become 37 and 122 respectively. In a similarly way we estimate that field multiplication (which has quadratic complexity) scales down to 309 cycles. A field inversion operation following an addition chain (1, 2, 4, 5, 10, 20, 40, 81, 162) requires nearly 22,700 cycles.

2. The for-loop in the scalar reduction operation (Algorithm 3) executes 163 times in $\mathbb{F}_{2^{163}}$ and performs linear operations such as additions/subtractions and shifting. Moreover the length of τ-adic representation of a scalar reduces to 163 (thus reducing by a factor of 0.57 in comparison to $\mathbb{F}_{2^{283}}$). So, we estimate that the cycle count for scalar conversion scales down by a factor of 0.57×0.61 and requires nearly 27,000 cycles.

3. One Frobenius-and-add operation over $\mathbb{F}_{2^{283}}$ in Algorithm 6 spends total 9,537 cycles among which 6,632 cycles are spent in eight quadratic-time field multiplications, and the rest 2,905 cycles are spent in linear-time operations. After scaling down, the cycle count for one Frobenius-and-add operation over $\mathbb{F}_{2^{163}}$ can be estimated to be around 4,250. The point multiplication loop iterates nearly 82 times for a τ-adic representation of length 164. Hence the number of cycles spent in this loop can be estimated to be around 348,500.

4. The precomputation and the final conversion steps are mainly dominated by the cost of field inversions. Hence the cycle counts can be estimated to be around 45,400.

As per the above estimates we see that a point multiplication using K-163 requires nearly 420,900 cycles. Similarly, we estimate that Montgomery's ladder for B-163 requires nearly 485,000 cycles. Our estimates show that our coprocessors for both B-163 and K-163 require more cycles in comparison to [34] which also uses a 16-

bit ALU. The reason behind this is that [34] uses a dual-port RAM, whereas our implementation uses a single-port RAM (as it works as a coprocessor of MSP430). Moreover [34] has a dedicated squarer circuit to minimize cycle requirement for squaring.

Table 3.1 provides estimates for cycle and area of a modified version of the coprocessor that performs point multiplications using the Montgomery's ladder on the NIST curve B-283. The estimated cycle count is calculated from the cycle counts of the field operations described in Sect. 3.4. From the estimated value, we see that a point multiplication on B-283 requires nearly 23.5% more time. However, the coprocessor for B-283 is smaller by around 550 GE as no scalar conversion is needed.

Although application-specific integrated circuits are the primary targets for our coprocessor, it may be useful also for FPGA-based implementations whenever small ECC designs are needed. Hence, we compiled our coprocessor also for Xilinx Spartan-6 XC6SLX4-2TQG144 FPGA by using Xilinx ISE 13.4 Design Suite. After place and route, it requires only 209 slices (634 LUTs and 309 registers) and runs on clock frequencies up to 106.598 MHz.

Our coprocessor significantly improves speed, both classical and side-channel security, memory, and energy consumption compared to leading lightweight software [3, 9, 16, 19, 30]. For example, [9] reports a highly optimized Assembly implementation running on a 32-bit Cortex-M0+ processor clocked at 48 MHz that computes a point multiplication on a less secure Koblitz curve K-233 without strong side-channel countermeasures. It computes a point multiplication in 59.18 ms (177.54 ms at 16 MHz) and consumes 34.16 μJ of energy.

3.6 Summary

In this chapter we showed that implementing point multiplication on a high security 283-bit Koblitz curve is feasible with extremely low resources making it possible for various lightweight applications. We also showed that Koblitz curves can be used in such applications even when the cryptosystem requires scalar conversions. Beside these contributions, we improved the scalar conversion by applying several optimizations and countermeasures against side-channel attacks. Finally, we designed a very lightweight architecture in only 4.3 kGE that can be used as a coprocessor for commercial 16-bit microcontrollers. Hence, we showed that Koblitz curves are feasible also for lightweight ECC even with on-the-fly scalar conversions and strong countermeasures against side-channel attacks.

References

1. Adikari J, Dimitrov VS, Järvinen K (2012) A fast hardware architecture for integer to τNAF conversion for koblitz curves. IEEE Trans Comput 61(5):732–737
2. Al-Daoud E, Mahmod R, Rushdan M, Kilicman A (2002) A new addition formula for elliptic curves over $GF(2^n)$. IEEE Trans Comput 51(8):972–975

3. Aranha DF, Dahab R, López J, Oliveira LB (2010) Efficient implementation of elliptic curve cryptography in wireless sensors. Adv Math Commun 4(2):169–187

4. Azarderakhsh R, Järvinen KU, Mozaffari-Kermani M (2014) Efficient algorithm and architecture for elliptic curve cryptography for extremely constrained secure applications. IEEE Trans Circuits Syst I Regul Pap 61(4):1144–1155

5. Batina L, Mentens N, Sakiyama K, Preneel B, Verbauwhede I (2006) Low-cost elliptic curve cryptography for wireless sensor networks. In: Security and privacy in ad-hoc and sensor networks — ESAS 2006. Lecture notes in computer science, vol 4357. Springer, Berlin, pp 6–17

6. Bock H, Braun M, Dichtl M, Hess E, Heyszl J, Kargl W, Koroschetz H, Meyer B, Seuschek H (2008) A milestone towards RFID products offering asymmetric authentication based on elliptic curve cryptography. In: Proceedings of the 4th workshop on RFID security — RFIDSec 2008

7. Brumley BB, Järvinen KU (2010) Conversion algorithms and implementations for koblitz curve cryptography. IEEE Trans Comput 59(1):81–92

8. Coron J-S (1999) Resistance against differential power analysis for elliptic curve cryptosystems. In: Cryptographic hardware and embedded systems — CHES 1999. Lecture notes in computer science, vol 1717. Springer, Berlin, pp 292–302

9. de Clercq R, Uhsadel L, Van Herrewege A, Verbauwhede I (2014) Ultra low-power implementation of ECC on the ARM cortex-M0+. In: Proceedings of the 51st annual design automation conference, DAC '14. ACM, New York, NY, USA, pp 112:1–112:6

10. Dimitrov VS, Järvinen KU, Jacobson MJ, Chan WF, Huang Z (2006) FPGA implementation of point multiplication on koblitz curves using kleinian integers. In: Cryptographic hardware and embedded systems, CHES'06. Springer, Berlin, pp 445–459

11. Dimitrov VS, Järvinen KU, Jacobson MJ, Chan WF, Huang Z (2008) Provably sublinear point multiplication on koblitz curves and its hardware implementation. IEEE Trans Comput 57:1469–1481

12. Fan J, Verbauwhede, I (2012) An updated survey on secure ECC implementations: attacks, countermeasures and cost. In: Cryptography and security: from theory to applications. Lecture notes in computer science, vol 6805. Springer, Berlin, pp 265–282

13. Fouque P-A, Valettem, F (2003) The doubling attack—why upwards is better than downwards. In: Cryptographic hardware and embedded systems — CHES 2003. Lecture notes in computer science, vol 2779. Springer, Berlin, pp 269–280

14. Hankerson D, Menezes AJ, Vanstone S (2003) Guide to elliptic curve cryptography. Springer, New York

15. Hein D, Wolkerstorfer J, Felber N (2009) ECC is ready for RFID–a proof in silicon. In: Selected areas in cryptography — SAC 2008. Lecture notes in computer science, vol 5381. Springer, Berlin, pp 401–413

16. Hinterwälder G, Moradi A, Hutter M, Schwabe P, Paar C (2015) Full-size high-security ECC implementation on MSP430 microcontrollers. In: Progress in cryptology — LATINCRYPT 2014. Lecture notes in computer science. Springer, Berlin, pp 31–47

17. Itoh T, Tsujii S (1988) A fast algorithm for computing multiplicative inverses in $GF(2^m)$ using normal bases. Inf Comput 78(3):171–177

18. Järvinen KU, Forsten J, Skyttä JO (2006) Efficient circuitry for computing τ-adic non-adjacent form. In: Proceedings of the IEEE international conference on electronics, circuits and systems (ICECS '06), pp 232–235

19. Kargl A, Pyka S, Seuschek H (2008) Fast arithmetic on ATmega128 for elliptic curve cryptography. Cryptology ePrint Archive, Report 2008/442

20. Koblitz N (1991) CM-curves with good cryptographic properties. In: Advances in cryptology — CRYPTO '91. Lecture notes in computer science, vol 576. Springer, Berlin, pp. 279–287

21. Kumar S, Paar C, Pelzl J, Pfeiffer G, Schimmler M (2006) Breaking ciphers with COPA-COBANA — a cost-optimized parallel code breaker. In: Cryptographic hardware and embedded systems (CHES 2006). Lecture notes in computer science, vol 4249. Springer, Berlin, pp 101–118

22. Lee YK, Sakiyama K, Batina L, Verbauwhede I (2008) Elliptic-curve-based security processor for RFID. IEEE Trans Comput 57(11):1514–1527
23. López J, Dahab R (1999) Improved algorithms for elliptic curve arithmetic in $GF(2^n)$. In: Selected areas in cryptography — SAC'98. Lecture notes in computer science, vol 1556. Springer, Berlin, pp 201–212
24. National Institute of Standard and Technology (2000) Federal information processing standards publication, FIPS 186–2. Digital Signature Standard
25. National Institute of Standards and Technology (2013) Digital signature standard (DSS). Federal information processing standard, FIPS PUB 186-4
26. Okeya K, Takagi T, Vuillaume C (2005) Efficient representations on koblitz curves with resistance to side channel attacks. In: Proceedings of the 10th Australasian conference on information security and privacy — ACISP 2005. Lecture notes in computer science, vol 3574. Springer, Berlin, pp 218–229
27. Pessl P, Hutter M (2014) Curved tags — a low-resource ECDSA implementation tailored for RFID. In: Workshop on RFID security — RFIDSec 2014
28. Schaumont PR (2013) A practical introduction to hardware/software codesign, 2nd edn. Springer, Berlin
29. Solinas JA (2000) Efficient arithmetic on koblitz curves. Des Codes Cryptogr 19(2–3):195–249
30. Szczechowiak P, Oliveira LB, Scott M, Collier M, Dahab R (2008) NanoECC: testing the limits of elliptic curve cryptography in sensor networks. In: European conference on wireless sensor networks — ESWN 2008. Lecture notes in computer science, vol 4913. Springer, Berlin, pp 305–320
31. Texas Instruments (2007–2012) MSP430F261x and MSP430F241x, June 2007, Revised November 2012. http://www.ti.com/lit/ds/symlink/msp430f2618.pdf Accessed 22 July 2015
32. Vuillaume C, Okeya K, Takagi T (2006) Defeating simple power analysis on koblitz curves. IEICE Trans Fundam Electron Commun Comput Sci E89-A(5):1362–1369
33. Wenger E (2013) Hardware architectures for MSP430-based wireless sensor nodes performing elliptic curve cryptography. In: Applied cryptography and network security — ACNS 2013. Lecture notes in computer science, vol 7954. Springer, Berlin, pp 290–306
34. Wenger E, Hutter M (2011) A hardware processor supporting elliptic curve cryptography for less than 9 kGEs. In: Smart card research and advanced applications — CARDIS 2011. Lecture notes in computer science, vol 7079. Springer, Berlin, pp 182–198

Chapter 4
Discrete Gaussian Sampling

CONTENT SOURCES:

Sujoy Sinha Roy, Frederik Vercauteren, Ingrid Verbauwhede High precision discrete Gaussian sampling on FPGAs In *International Conference on Selected Areas in Cryptography SAC* (2013).

Contribution: Main author.

Sujoy Sinha Roy, Oscar Reparaz, Frederik Vercauteren, Ingrid Verbauwhede Compact and Side Channel Secure Discrete Gaussian Sampling In *IACR Cryptology ePrint Archive eprint/2014/591* (2014).

Contribution: Main author.

4.1 Introduction

In this chapter we propose an efficient hardware implementation of a discrete Gaussian sampler for ring-LWE encryption schemes. The proposed sampler architecture is based on the Knuth-Yao sampling Algorithm [10]. It has high precision and large tail-bound to keep the statistical distance below 2^{-90} to the true Gaussian distribution for the secure parameter sets [6] that are used in the public key encryption schemes [12, 17].

The remaining part of the chapter is organized as follows: In Sect. 4.2 we describe the Knuth-Yao sampling algorithm in detail. In the next section we analyze the Knuth-Yao algorithm and design an efficient algorithm that consumes very little amount of resources. The hardware architecture of the discrete Gaussian sampler is presented in Sect. 4.4. In Sect. 4.5 we describe side channel vulnerability of the sampler architecture along with countermeasures. Detailed experimental results are presented in Sect. 4.6. The final section has the conclusion.

© Springer Nature Singapore Pte Ltd. 2020
S. Sinha Roy and I. Verbauwhede, *Lattice-Based Public-Key Cryptography in Hardware*,
Computer Architecture and Design Methodologies,
https://doi.org/10.1007/978-981-32-9994-8_4

Fig. 4.1 Probability matrix
and corresponding DDG-tree

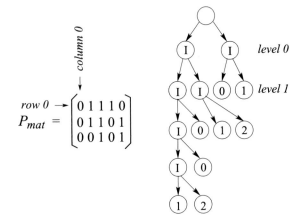

$$\begin{array}{c} \textit{row 0} \rightarrow \\ P_{mat} = \end{array} \begin{bmatrix} 0\ 1\ 1\ 1\ 0 \\ 0\ 1\ 1\ 0\ 1 \\ 0\ 0\ 1\ 0\ 1 \end{bmatrix}$$

4.2   The Knuth-Yao Algorithm

The Knuth-Yao algorithm uses a random walk model to perform sampling using the probabilities of the sample space elements. The method is applicable for any non-uniform distribution. Let p_j be the probability of the jth sample in the sample space. The binary expansions of the probabilities of the samples are written in the form of a matrix which we call the *probability matrix* P_{mat}. The jth row of the probability matrix corresponds to the binary expansion of p_j. An example of the probability matrix for a sample space containing three sample points $\{0, 1, 2\}$ with probabilities $p_0 = 0.01110$, $p_1 = 0.01101$ and $p_2 = 0.00101$ is shown in Fig. 4.1.

A rooted binary tree known as a discrete distribution generating (DDG) tree is constructed from the probability matrix. Each level of the DDG tree can have two types of nodes: intermediate nodes (I) and terminal nodes. The number of terminal nodes in the ith level of the DDG tree is equal to the Hamming weight of ith column in the probability matrix. Here we provide an example of the DDG tree construction for the given probability distribution in Fig. 4.1. The root of the DDG tree has two children which form the 0th level. Both the nodes in this level are marked with I since the 0th column in P_{mat} does not contain any non-zero. These two intermediate nodes have four children in the 1st level. To determine the type of the nodes, the 1st column of P_{mat} is scanned from the bottom. In this column only the row numbers '1' and '0' are non-zero; hence the right-most two nodes in the 1st level of the tree are marked with '1' and '0' respectively. The remaining two nodes in this level are thus marked as intermediate nodes. Similarly the next levels are also constructed. The DDG tree corresponding to P_{mat} is given in Fig. 4.1. At any level of the DDG tree, the terminal nodes (if present) are always on the right hand side.

The sampling operation is a random walk which starts from the root; visits a left-child or a right-child of an intermediate node depending on the random input bit. The sampling process completes when the random walk hits a terminal node and the output of the sampling operation is the value of the terminal node. By construction, the Knuth-Yao random walk samples accurately from the distribution defined by the probability matrix.

The space requirement for the DDG tree can be reduced by constructing it at run time during the sampling process. As shown in Fig. 4.1, the ith level of the DDG tree is completely determined by the $(i-1)$th level and the ith column of the probability matrix. Hence it is sufficient to store only one level of the DDG tree during the sampling operation and construct the next level on the fly (if required) using the probability matrix [10].

In fact, in the next section we introduce a novel method to traverse the DDG tree that only requires the current node and the ith column of the probability matrix to derive the next node in the tree traversal.

4.3   DDG Tree on the Fly

In this section we propose an efficient hardware-implementation of the Knuth-Yao based discrete Gaussian sampler which samples with high precision and large tail-bound. We describe how the DDG tree can be traversed efficiently in hardware and then propose an efficient way to store the probability matrix such that it can be scanned efficiently and also requires near-optimal space. Before we describe the implementation of the sampler, we first recall the parameter set for the discrete Gaussian sampler from the LWE implementation in [6].

4.3.1   Parameter Sets for the Discrete Gaussian Sampler

Table 4.1 shows the tail bound $|z_t|$ and precision ϵ required to obtain a statistical distance of less than 2^{-90} for the Gaussian distribution parameters in Table 4.1 of [6]. The dimension of the lattice is m. The standard deviation σ in Table 4.1 is derived from the parameter s using the equation $s = \sigma\sqrt{2\pi}$. The tail bound $|z_t|$ is calculated from Eq. 2.7 for the right-hand upper bound 2^{-100}. For a maximum statistical distance of 2^{-90} and a tail bound $|z_t|$, the required precision ϵ is calculated using Eq. 2.8.

However in practice the tail bounds are quite loose for the precision values in Table 4.1. The probabilities are zero (upto the mentioned precision) for the sample points greater than 39 for all three distributions. Given a probability distribution, the Knuth-Yao random walk always hits a sample point when the sum of the probabilities is one [10]. However if the sum is less than one, then the random walk may not hit a terminal node in the corresponding DDG tree. Due to finite range and precision in Table 4.1, the sum of the discrete Gaussian probability expansions (say P_{sum}) is less than one. We take a difference $(1 - P_{sum})$ as another sample point which indicates *out of range* event. If the Knuth-Yao random walk hits this sample point, the sample value is discarded. This *out of range* event has probability less than 2^{-100} for all three distribution sets.

Table 4.1 Parameter sets to achieve statistical distance less than 2^{-90}

m	s	σ	Tail cut $|z_t|$	Precision ϵ
256	8.35	3.33	84	106
320	8.00	3.192	86	106
512	8.01	3.195	101	107
1024	8.01	3.195	130	109

4.3.2   Construction of the DDG Tree During Sampling

During the Knuth-Yao random walk, the DDG tree is constructed at run time. The implementation of DDG tree as a binary tree data structure is an easy problem [1] in software, but challenging on hardware platforms. As described in Sect. 4.2, the implementation of the DDG tree requires only one level of the DDG tree to be stored. However the ith level of a DDG tree may contain as many as 2^i nodes. On software platforms, dynamic memory allocation can be used at run time to allocate sufficient memory required to store a level of the DDG tree. But in hardware, we need to design the sampler architecture for the worst case storage requirement which makes the implementation costly.

We propose a hardware implementation friendly traversal based on specific properties of the DDG tree. We observe that in a DDG tree, all the intermediate nodes are on the left hand side; while all the terminal nodes are on the right hand side. This observation is used to derive a simple algorithm which identifies the nodes in the DDG tree traversal path instead of constructing each level during the random walk. Figure 4.2 describes the $(i-1)$th level of the DDG tree. The intermediate nodes are I, while the terminal nodes are T. The node visited at this level during the sampling process is highlighted by the double circle in the figure. Assume that the visited node is not a terminal node. This assumption is obvious because if the visited node is a terminal node, then we do not need to construct the ith level of the DDG tree. At this level, let there be n intermediate nodes and the visited node is the kth node from the left. Let $d = n - k$ denote the distance of the right most intermediate node from the visited node.

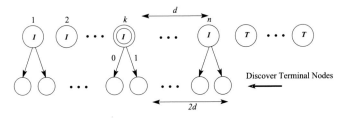

Fig. 4.2 DDG tree construction

In the next step, the sampling algorithm reads a random bit and visits a child node on the ith level of the DDG tree. If the visited node is a left child, then it has $2d + 1$ nodes to its right side. Otherwise, it will have $2d$ nodes to its right side (as shown in the figure). To determine whether the visited node is a terminal node or an intermediate node, the ith column of the probability matrix is scanned. The scanning process detects the terminal nodes from the right side of the ith level and the number of terminal nodes is equal to the Hamming weight h of the ith column of the probability matrix. The left child is a terminal node if $h > (2d + 1)$ and the right child is a terminal node if $h > 2d$. If the visited node is a terminal node, we output the corresponding row number in the probability matrix as the result of sampling process. When the visited node in the ith level is internal, its visited-child in the $(i + 1)$th level is checked in a similar way.

From the analysis of DDG tree construction, we see the following points :

1. The sampling process is independent of the internal nodes that are to the left of the visited node.
2. The terminal nodes on the $(i - 1)$th level have no influence on the construction of the ith level of the DDG tree.
3. The distance d between the right most internal node and the visited node on the $(i - 1)$th level of the DDG tree is sufficient (along with the Hamming weight of the ith column of the probability matrix) to determine whether the visited node on the ith level is an internal node or a terminal node.

During the Knuth-Yao sampling we do not store an entire level of the DDG tree. Instead, the difference d between the visited node and the right-most intermediate

---

Algorithm 7: *Knuth-Yao Sampling*

Input: Probability matrix P
Output: Sample value S

```
 1 begin
 2 │   d ← 0; /* Distance between the visited and the rightmost internal node */
 3 │   Hit ← 0; /* This is 1 when the sampling process hits a terminal node */
 4 │   col ← 0;     /* Column number of the probability matrix */
 5 │   while Hit = 0 do
 6 │   │   r ← RandomBit() ;
 7 │   │   d ← 2d + r̄ ;
 8 │   │   for row = MAXROW down to 0 do
 9 │   │   │   d ← d − P[row][col] ;
10 │   │   │   if d = −1 then
11 │   │   │   │   S ← row ;
12 │   │   │   │   Hit ← 1 ;
13 │   │   │   │   ExitForLoop() ;
14 │   │   │   end
15 │   │   end
16 │   │   col ← col + 1 ;
17 │   end
18 end
```

---

node is used to construct the visited node at the next level. The steps of the Knuth-Yao sampling operation are described in Algorithm. 7. In Line 6, a random bit r is used to jump to the next level of the DDG tree. On this new level, the distance between the visited node and the rightmost node is initialized to either $2d$ or $2d + 1$ depending on the random bit r. In Line 8, the for-loop scans a column of the probability matrix to detect the terminal nodes. Whenever the algorithm finds a 1 in the column, it detects a terminal node. Hence, the relative distance between the visited node and the right most internal node is decreased by one (Line 9). When d is reduced to -1, the sampling algorithm hits a terminal node. Hence, in this case the sampling algorithm stops and returns the corresponding row number as the output. In the other case, when d is positive after completing the scanning of an entire column of the probability matrix, the sampling algorithm jumps to the next level of the DDG tree.

4.3.3 Storing the Probability Matrix Efficiently

The Knuth-Yao algorithm reads the probability matrix of the discrete Gaussian distribution during formation of the DDG tree. A probability matrix having r rows and c columns requires rc bits of storage. This storage could be significant when both r (depends on the tail-bound) and c (depends on the precision) are large. As an example, Fig. 4.3 shows a portion of the probability matrix for the probabilities of $0 \le |z| \le 17$ with 30-bits precision according to the discrete Gaussian distribution with parameter $s = 8.01$. In [5] the authors observed that the leading zeros in the probability matrix can be compressed. The authors partitioned the probability matrix in different blocks having equal (or near-equal) number of leading zeros. Now for any row of the probability matrix, the conditional probability with respect to the block it belongs to is calculated and stored. In this case the conditional probability expansions

Fig. 4.3 Storing probability matrix

```
000111111111010111000101110101
001111001101110110011011001101
001101001000110011101100011010
001010010010001110000011001110
000111010011001101100110100000
000100101100101100100011010010
000010101111011110010010001110
000001011100110110001001011000
000001011001000101100101011101
000000011001101100000011010010
000000001111010010001111111011
000000000101011011011001001
000000000011100010111000110
000000000001000010101101010101
000000000000010001110010001
000000000000001000100110001
000000000000000001111000100
000000000000000000010111111
```

Part of Probability Matrix   First two ROM words

#2

#1 | 11011_110010111_11

#0 | 001110_1110111_110

do not contain a long sequence of leading zeros. The precision of the conditional probabilities is less than the precision of the absolute probabilities by roughly the number of leading zeros present in the absolute probability expansions. The sampling of [5] then applies two rounds of the Knuth-Yao algorithm: first to select a block and then to select a sample value according to the conditional probability expansions within the block.

However the authors of [5] do not give any actual implementation details. In hardware, ROM is ideal for storing a large amount of fixed data. To minimize computation time, data fetching from ROM should be minimized as much as possible. The pattern in which the probability expansions are stored in ROM determines the number of ROM accesses (thus performance) during the sampling process. During the sampling process the probability matrix is scanned column by column. Hence to ease the scanning operation, the probability expansions should be stored in a column-wise manner in ROM.

In Fig. 4.3, the probability matrix for a discrete Gaussian distribution contains large chunks of zeros near the bottom of the columns. Since we store the probability matrix in a column-wise manner in ROM, we perform compression of zeros present in the columns. The *column length* is the length of the top portion after which the chunk of bottom zeros start. We target to optimize the storage requirement by storing only the upper portions of the columns in ROM. Since the columns have different lengths, we also store the lengths of the columns. The number of bits required to represent the length of a column can be reduced by storing only the difference in column length with respect to the previous column. In this case, the number of bits required to represent the differential column length is the number of bits in the maximum deviation and a sign bit. For the discrete Gaussian distribution matrix shown in Fig. 4.3, the maximum deviation is three and hence three bits are required to represent the differential column lengths. Hence the total number of bits required to store the differential column lengths of the matrix (Fig. 4.3) is 86 (ignoring the first two columns).

For the discrete Gaussian distribution matrix, we observe that the difference between two consecutive column lengths is one for most of the columns. This observation is used to store the distribution matrix more efficiently in ROM. We consider only non-negative differences between consecutive column lengths; the length of a column either increases or remains the same with respect to its left column. When there is a decrement in the column length, the extra zeros are also considered to be part of the column to keep the column length the same as its left neighbor. In Fig. 4.3 the dotted line is used to indicate the lengths of the columns. It can be seen that the maximum increment in the column length happens to be one between any two consecutive columns (except the initial few columns). In this representation only one bit per column is needed to indicate the difference with respect to the left neighboring column: 0 for no-increment and 1 for an increment by one. With such a representation, 28 bits are required to represent the increment of the column lengths for the matrix in Fig. 4.3. Additionally, 8 redundant zeros are stored at the bottom of the columns due to the decrease in column length in a few columns. Thus, a total of 36 bits are stored in addition to the pruned probability matrix. There is one more

---

Algorithm 8: *Knuth-Yao Sampling in Hardware Platform*

Input: Probability matrix P
Output: Sample value S

1 **begin**
2 $\quad$ $d \leftarrow 0;$ $\quad$ /* Distance between the visited and the rightmost internal node */
3 $\quad$ $Hit \leftarrow 0;$ $\quad$ /* This is 1 when the sampling process hits a terminal node */
4 $\quad$ $ColLen \leftarrow INITIAL;$ $\quad$ /* Column length is set to the length of first column */
5 $\quad$ $address \leftarrow 0;$ $\quad$ /* This variable is the address of a ROM word */
6 $\quad$ $i \leftarrow 0;$ $\quad$ /* This variable points the bits in a ROM word */
7 $\quad$ **while** $Hit = 0$ **do**
8 $\quad\quad$ $r \leftarrow RandomBit();$
9 $\quad\quad$ $d \leftarrow 2d + \bar{r};$
10 $\quad\quad$ $ColLen \leftarrow ColLen + ROM[address][i];$
11 $\quad\quad$ **for** $row = ColLen - 1$ *down to* 0 **do**
12 $\quad\quad\quad$ $i \leftarrow i + 1;$
13 $\quad\quad\quad$ **if** $i = w$ **then**
14 $\quad\quad\quad\quad$ $address \leftarrow address + 1;$
15 $\quad\quad\quad\quad$ $i \leftarrow 0;$
16 $\quad\quad\quad$ **end**
17 $\quad\quad\quad$ $d \leftarrow d - ROM[row][i];$
18 $\quad\quad\quad$ **if** $d = -1$ **then**
19 $\quad\quad\quad\quad$ $S \leftarrow row;$
20 $\quad\quad\quad\quad$ $Hit \leftarrow 1;$
21 $\quad\quad\quad\quad$ $ExitForLoop();$
22 $\quad\quad\quad$ **end**
23 $\quad\quad$ **end**
24 $\quad$ **end**
25 $\quad$ **return** (S)
26 **end**

---

advantage of storing the probability matrix in this way in that we can use a simple binary counter to represent the length of the columns. The binary counter increments by one or remains the same depending on the column-length increment bit.

In ROM, we only store the portion of a column above the partition-line in Fig. 4.3 along with the column length difference bit. The column-length difference bit is kept at the beginning and then the column is kept in reverse order (bottom-to-top). As the Knuth-Yao algorithm scans a column from bottom to top, the column is stored in reverse order. Figure 4.3 shows how the columns are stored in the first two ROM words (word size 16 bits). During the sampling process, a variable is used to keep track of the column-lengths. This variable is initialized to the length of the first non-zero column. For the probability matrix in Fig. 4.3, the initialization value is 5 instead of 4 as the length of the next column is 6. Whilst scanning a new column, this variable is either incremented (starting bit 1) or kept the same (starting bit 0). Algorithm 8 summarizes the steps when a ROM of word size w is used as a storage for the probability matrix.

4.3.4 Fast Sampling Using a Lookup Table

A Gaussian distribution is concentrated around its center. In the case of a discrete Gaussian distribution with standard deviation σ, the probability of sampling a value larger than $t \cdot \sigma$ is less than $2 \exp(-t^2/2)$ [14]. In fact this upper bound is not very tight. We use this property of a discrete Gaussian distribution to design a fast sampler architecture satisfying the speed constraints of many real-time applications. As seen from the previous section, the Knuth-Yao random walk uses random bits to move from one level of the DDG tree to the next level. Hence the average case computation time required per sampling operation is determined by the number of random bits required in the average case.

The lower bound on the number of random bits required per sampling operation in the average case is given by the entropy of the probability distribution [3]. The entropy of a continuous normal distribution with a standard deviation σ is $\frac{1}{2} \log(2\pi e\sigma^2)$. For a discrete Gaussian distribution, the entropy is approximately close to entropy of the normal distribution with the same standard deviation. A more accurate entropy can be computed from the probability values as per the following equation.

$$ H = - \sum_{-\infty}^{\infty} p_i \log p_i \ . \tag{4.1} $$

The Knuth-Yao sampling algorithm was developed to consume the minimum number of random bits on average [10]. It was shown that the sampling algorithm requires at most $H + 2$ random bits per sampling operation in the average case.

For a Gaussian distribution, the entropy H increases with the standard deviation σ, and thus the number of random bits required in the average case also increases with σ. For applications such as the ring-LWE based public key encryption scheme and homomorphic encryption, small σ is used. Hence for such applications the number of random bits required in the average case are small. Based on this observation we can avoid the costly bit-scanning operation using a small precomputed table that directly maps the initial random bits into a sample value (with large probability) or into an intermediate node in the DDG tree (with small probability). During a sampling operation, first a table lookup operation is performed using the initial random bits. If the table lookup operation returns a sample value, then the sampling algorithm terminates. For the other case, bit scanning operation is initiated from the intermediate node. For example, when $\sigma = 3.33$, if we use a precomputed table that maps the first eight random bits, then the probability of getting a sample value after the table lookup is 0.973. Hence using the lookup table we can avoid the costly bit-scanning operation with probability 0.973. However extra storage space is required for this lookup table. When the probability distribution is fixed, the lookup table can be implemented as a ROM which is cheap in terms of area in hardware platforms. In the next section we propose a cost effective implementation of a fast Knuth-Yao sampler architecture.

4.4   The Sampler Architecture

The sampler architecture is composed of (1) a bit-scanning unit, (2) counters for
column length and row number, and (3) a subtraction-based down counter for the
Knuth-Yao distance in the DDG tree. In addition, for the fast sampler architecture,
a lookup table is also used. A control unit is used to generate control signals for
the different blocks and to maintain synchronization between the blocks. We now
describe the different components of the sampler architecture.

4.4.1   The Bit-Scanning Unit

The bit-scanning unit is composed of a ROM, a scan register, one ROM-address
counter, one counter to record the number of bits scanned from a ROM-word and
a comparator. The ROM contains the probabilities and is addressed by the ROM-
address counter. During a bit-scanning operation, a ROM-word (size w bits) is first
fetched and then stored in the scan register. The scan-register is a shift-register and its
msb is read as the probability-bit. To count the number of bits scanned from a ROM-
word, a counter *word-bit* is used. When the *word-bit* counter reaches $w-2$ from zero,
the output from the comparator *Comp1* enables the *ROM-address* counter. In the next
cycle the *ROM-address* counter addresses the next ROM-word. Also in this cycle
the *word-bit* counter reaches $w-1$ and the output from *Comp2* enables reloading
of the bit-scan register with the new ROM-word. In the next cycle, the *word-bit*
counter is reset to zero and the bit-scan register contains the word addressed by the
ROM-word counter. In this way data loading and shifting in the bit-scan register takes
place without any loss of cycles. Thus the frequency of the data loading operation
(which depends on the widths of the ROM) does influence the cycle requirement of
the sampler architecture. This interesting feature of the bit-scan unit will be utilized
in the next part of this section to achieve optimal area requirement by adjusting
the width of the ROM and the bit-scan register. The bit-scanning unit is the largest
sub-block in the sampler architecture in terms of area. Hence this unit should be
designed carefully to achieve minimum area requirement. In FPGAs a ROM can be
implemented as a distributed ROM or as a block RAM. When the amount of data is
small, a distributed ROM is the ideal choice. The way a ROM is implemented (its
width w and depth h) affects the area requirement of the sampler. Let us assume that
the total number of probability bits to be stored in the ROM is D and the size of the
FPGA LUTs is t. Then the total number of LUTs required by the ROM is around
$\lceil \frac{D}{w \cdot 2^t} \rceil \cdot w$ along with a small amount of addressing overhead. The scan-register is
a shift register of width w and consumes around w LUTs and $w_f = w$ FFs. Hence
the total area (LUTs and FFs) required by the ROM and the scan-register can be
approximated by the following equation.

$$\#Area = \left\lceil \frac{D}{w \cdot 2^t} \right\rceil \cdot w + (w + w_f) = \left\lceil \frac{h}{2^t} \right\rceil \cdot w + (w + w_f) .$$

Table 4.2 Area of the bit-scan unit for different widths and depths

Width	Height	LUTs	FFs	Slices
24	128	70	35	22
12	256	72	23	18
6	512	67	17	18

For optimal storage, h should be a multiple of 2^t. Choosing a larger value of h will reduce the width of the ROM and hence the width of the scan-register. However with the increase in h, the addressing overhead of the ROM will also increase. In Table 4.2 we compare area of the bit-scan unit for $\sigma = 3.33$ with various widths of the ROM and the scan register using Xilinx Virtex V xcvlx30 FPGA. The optimal implementation is achieved when the width of the ROM is set to six bits. Though the slice count of the bit-scan unit remains the same in both the second and third column of the table due to various optimizations performed by the Xilinx ISE tool, the actual effect on the overall sampler architecture will be evident in Sect. 4.6.

4.4.2 Row-Number and Column-Length Counters

As described in the previous section, we use a one-step differential encoding for the column lengths in the probability matrix. The *column-length* counter in Fig. 4.4 is an up-counter and is used to represent the lengths of the columns. During a random-walk, this counter increments depending on the column-length bit which appears in the starting of a column. If the column-length bit is zero, then the *column-length* counter remains in its previous value; otherwise it increments by one. At the starting of a column-scanning operation, the *Row-number* counter is first initialized to the value of column-length. During the scanning operation this counter decrements by one in each cycle. A column is completely read when the *Row Number* counter reaches zero.

4.4.3 The Distance Counter

A subtraction-based counter *distance* is used to keep the distance d between the visited node and the right-most intermediate node in the DDG tree. The register *distance* is first initialized to zero. During each column jump, the *row_zero_reg* is set and thus the subtrahend becomes zero. In this step, the *distance* register is updated with the value $2d$ or $2d + 1$ depending on the input random bit. As described in the previous section, a terminal node is visited by the random walk when the distance becomes negative for the first time. This event is detected by the control FSM using the carry generated from the subtraction operation.

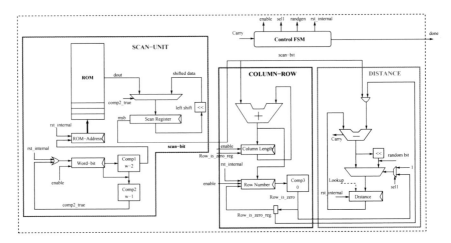

Fig. 4.4 Hardware architecture for Knuth-Yao sampler

After completion of a random walk, the value present in *Row Number* is the magnitude of the sample output. One random bit is used as a sign of the value of the sample output.

4.4.4 The Lookup Table for Fast Sampling

The output from the Knuth-Yao sampling algorithm is determined by the probability distribution and by the input sequence of random bits. For a given fixed probability distribution, we can precompute a table that maps all possible random strings of bit-width s into a sample point or into an intermediate distance in the DDG tree. The precomputed table consists of 2^s entries for each of the 2^s possible random numbers.

On FPGAs, this precomputed table is implemented as a distributed ROM using LUTs. The ROM contains 2^s words and is addressed by random numbers of s bit width. The success probability of a table lookup operation can be increased by increasing the size of the lookup table. For example when $\sigma = 3.33$, the probability of success is 0.973 when the lookup table maps the eight random bits; whereas the success probability increases to 0.999 when the lookup table maps 13 random bits. However with a larger mapping, the size of precomputed table increases exponentially from 2^8 to 2^{13}. Additionally each lookup operation requires 13 random bits. A more efficient approach is to perform lookup operations in steps. For example, we use a first lookup table that maps the first eight random bits into a sample point or an intermediate distance (three bit wide for $\sigma = 3.33$). In case of a lookup failure, the next step of the random walk from the obtained intermediate distance will be determined by the next sequence of random bits. Hence, we can extend the lookup operation to speedup the sampling operation. For example, the three-bit wide

Fig. 4.5 Hardware architecture for two stage Lookup

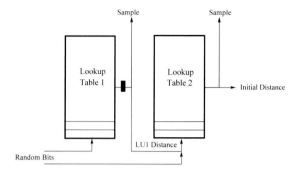

distance can be combined with another five random bits to address a (the *second*) lookup table. Using this two small lookup tables, we achieve a success probability of 0.999 for $\sigma = 3.33$. An architecture for a two stage lookup table is shown in Fig. 4.5.

4.5 Timing and Simple Power Analysis

The Knuth-Yao sampler presented in this chapter is not a constant-time architecture as it uses a bit scanning operation in which the sample generated is related to the number of probability-bits scanned during a sampling operation. Hence, the number of cycles for a sampling operation provides some information about the sample. In 2016, Bruinderink et al. [7] mounted an attack on the BLISS [4] signature scheme by exploiting the timing information leakage from the non constant-time discrete Gaussian sampler. In particular, the attack targets a software implementation of the BLISS scheme and exploits cache weakness of the Gaussian sampler to run the LLL lattice reduction [11] algorithm. The attack requires as little as 450 signatures to recover the secret key. The attack technique is described in more details in [7].

We recall that in the ring-LWE encryption scheme in Sect. 2.4.1, the Gaussian sampler is used during key generation and message encryption. The key generation operation is performed only to generate long-term keys and hence can be performed in a secure environment. However, this is not the case for the encryption operation where an encoded message $\bar{m}$ is masked as $c_2 = p \cdot e_1 + e_3 + \bar{m}$ using two Gaussian distributed error polynomials e_1 and e_3. It should be noted that in a public key encryption scheme, the plaintext is normally considered secret information. For example, it is a common practice to use a public-key cryptosystem to encrypt a symmetric key that is subsequently used for fast, bulk encryption (this construction is commonly named "hybrid cryptosystems"). Hence, from the perspective of side-channel analysis, any leak of information during the encryption operation about the plaintext (symmetric key) is considered as a valid security threat. We would like to mention that the timing attack proposed by Bruinderink et al. [7] on the BLISS signature scheme may not be applied directly on the ring-LWE public key encryption scheme as the attacker can

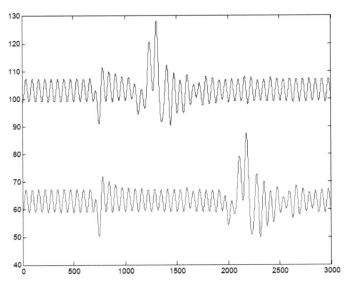

Fig. 4.6 Two instantaneous power consumption measurements corresponding to two different sampling operations. Horizontal axis is time, vertical axis is electromagnetic field intensity. The different timing for the two different sampling operations is evident

have only one ciphertext per message. Whereas for the signature scheme, the attacker can use the LLL algorithm since she can have several signatures for the same secret key.

To verify to what extent the instantaneous power consumption provides information about the sampling operation, we performed an SPA attack on the unprotected design running on a Xilinx Spartan-III at 40 MHz. The instantaneous power consumption is measured with a Langer RF5-2 magnetic pick-up coil on top of the FPGA package (without decapsulation), amplified (+50 dB), low-pass filtered (cut-off frequency of 48 MHz). In Fig. 4.6 we show the instantaneous power consumption of two different sampling operations. The horizontal axis denotes time, and both sampling operations are triggered on the beginning of the sampling operation. One can distinguish enough SPA features (presumably due to register updates) to infer that the *blue* graph corresponds to a sampling that requires small number of cycles (7 cycles exactly) whereas the *red* graph represents a sampling operation that requires more cycles (21 cycles). With this SPA attack, the adversary can predict the *magnitudes* of the Gaussian distributed samples. Note that this information is partial as the value of a sample also includes a random sign bit. We remark that with a more sophisticated side channel attack, it might be possible to recover the sign bits by observing the modular arithmetic operations during the message encryption. If possible, then the message can be extracted easily from the ciphertext.

4.5.1 Strategies to Mitigate the Side-Channel Leakage

In this chapter we propose an efficient and cost effective scheme to protect the Gaussian sampler from simple timing and power analysis based attacks. Our proposal is based on the fact that the encryption scheme remains secure as long as the attacker has no information about the relative positions of the samples (i.e. the coefficients) in the noise polynomials. It should be noted that, as the Gaussian distribution used in the encryption scheme is a publicly known parameter, any one can guess the number of a particular sample point in an array of n samples. Similar arguments also apply for other cryptosystems where the key is a uniformly distributed random string of bits of some length (say l). For such a random key, one has the information that almost half of the bits in the key are one and the rest are zero. In other words, the Hamming weight is around $l/2$. Even if the exact value of the Hamming weight is revealed to the adversary (on average, say $l/2$), the key still mantains $\log_2 \binom{l}{l/2}$ bits of entropy (≈ 124 bits for a 128 bit key). It is the random positions of the bits that make a key secure.

In the ring-LWE encryption scheme (Sect. 2.4.1), the Gaussian sampler is used to generate error polynomials. The sequential bit scanning operation reveals information about the samples and their positions in the error polynomials. Our strategy against simple timing and power analysis attack is described below:

1. Use of a lookup: The table lookup operation is constant-time and has a very large success probability. Hence with this lookup approach, we protect most of the samples from leaking any information about the value of the sample from which an attacker can perform simple power and timing analysis.
2. Use of a random permutation: The table lookup operation succeeds in most events, but fails with a small probability. For a failure, the sequential bit scanning operation leaks information about the samples. For example, when $\sigma = 3.33$ and the lookup table maps initial eight random bits, the bit scanning operation is required for seven samples out of 256 samples in the average case. To protect against SPA, we perform a random shuffle after generating an entire array of samples. The random shuffle operation swaps all bit-scan operation generated samples with other random samples in the array. This random shuffling operation removes any timing information which an attacker can exploit. In the next section we will describe an efficient implementation of the random shuffling operation.

4.5.2 Efficient Implementation of the Random Shuffling

We use a modified version of the Fisher and Yates shuffle which is also known as the *Knuth shuffle* [9] to perform random shuffling of the bit-scan operation generated samples. The advantages of this shuffling algorithm are its simplicity, uniformness, inplace data handling and linear time complexity. In the original shuffling algorithm, all the indexes of the input array are processed one after another. However in our case we can restrict the shuffling operation to only those samples that were generated using the sequential bit scanning operation. This operation is implemented in the following way.

Assume that m samples are generated and then stored in a RAM with addresses in the range 0 to $(m - 1)$. We use two counters C_1 and C_2 to represent the number of samples generated through successful lookup and bit-scanning operations respectively. The total number of samples generated is given by $(C_1 + C_2)$. The samples generated using lookup operation are stored in the memory locations starting from 0 till $(C_1 - 1)$; whereas the bit-scan generated samples are stored in the memory locations starting from address $m - 1$ downto $m - C_2$. After generation of the m samples, the bit-scan operation generated samples are randomly swapped with the other samples using Algorithm 9.

Algorithm 9: *Random swap of samples*

Input: Sample vector stored in RAM[ ] with timing information
Output: Sample vector stored in RAM[ ] without timing information
1 **begin**
2   **while** $C_2 > 0$ **do**
3       $L1 : random\_index \leftarrow random()$;
4   **if** $random\_index \geq (m - C_2)$ **then**
5   goto $L1$;
6   **end**
7   swap $RAM[m - C_2] \leftrightarrow RAM[random\_index]$;
8       $C_2 \leftarrow C_2 - 1$;
9   **end**
10 **end**

A hardware architecture for the secure consecutive-sampling is shown in Fig. 4.7. In the architecture, C_1 is an up-counter and C_2 is an up-down-counter. When the *enable* signal is high, the Gaussian sampler generates samples in an iterative way. After generation of each sample, the signal *Gdone* goes high and the type of the sample is indicated by the signal *lookup_success*. In the case when the sample has been generated using a successful lookup operation, *lookup_success* becomes high. Depending on the value of the *lookup_success*, the control machine stores the sample in the memory address C_1 or $(m - C_2)$ and also increments the corresponding counter. Completion of the m sampling operations is indicated by the output from *Comparator2*.

In the random-shuffling phase, a random address is generated and then compared with $(m - C_2)$. If the random-address is smaller than $(m - C_2)$ then it is used for the swap operation; otherwise another random-address is generated. Now the memory content of address $(m - C_2)$ is swapped with the memory content of random-address using the *ram_buffer* register. After this swap operation, the counter C_2 decrements by one. The last swap operation happens when C_2 is zero.

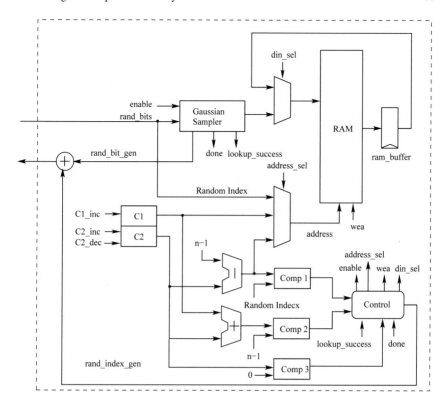

Fig. 4.7 Sampler with shuffling

Table 4.3 Performance of the discrete Gaussian sampler on xc5vlx30

Sampler architecture	ROM	LU	Area	Delay	Cycles
	Width/depth	Depth	LUTs/FFs/Slices/BRAM	ns	
Basic sampler	24/128	–	101/81/38/-	2.9	17
Basic sampler	12/256	–	105/60/32/-	2.5	17
Basic sampler★	6/512	–	102/48/30/-	2.6	17
Fast sampler	6/512	8	118/48/35/-	3	≈2.5
Bernoulli [16]	–/–	–	132/40/37/-	7.3	144
Polynomial sampler–1	6/512	8	135/56/44/1	3.1	392
Polynomial sampler–2	6/512	8	176/66/52/1	3.3	420

4.6 Experimental Results

We have evaluated the Knuth-Yao discrete Gaussian sampler architecture for $\sigma = 3.33$ using the Xilinx Virtex V FPGA xcvlx30 with speed grade -3. The results shown in Table 4.3 are obtained from the Xilinx ISE12.2 tool after place and route analysis. In the table we show area and timing results of our architecture for various configurations and modes of operations and compare the results with other existing architectures. The results do not include the area of the random bit generator. Area requirements for the basic bit-scan operation based Knuth-Yao sampler for different ROM-widths and depths are shown in the first three rows of the table. The optimal area is achieved when the ROM-width is set to 6 bits. As the width of the ROM does not affect the cycle requirement of the sampler architecture, all different configurations have same clock cycle requirement. The average case cycle requirement of the sampler is determined by the number of bits scanned on average per sampling operation. A C program simulation shows that the number of memory-bits scanned on average is 13.5. Before starting the bit-scanning operation, the sampler performs two column jump operations for the first two all-zero columns of the probability matrix (for $\sigma = 3.33$). This initial operation requires two cycles. After this, the bit scan operation requires 14 cycles to scan 14 memory-bits and the final transition to the completion state of the FSM requires one cycle. Thus, on average 17 cycles are spent per sampling operation. The compact Bernoulli sampler proposed in [16] consumes 37 slices and spends on average 144 cycles to generate a sample point.

The fast sampler architecture in the fourth column of Table 4.3 uses a lookup table that maps eight random bits. The sampler consumes additional five slices compared to the basic bit-scan based architecture. The probability that a table lookup operation returns a sample is 0.973. Due to this high success rate of the lookup operation, the average case cycle requirement of the fast sampler is slightly larger than 2 cycles with the consideration that one cycle is consumed for the transition of the state-machine to the completion state. In this cycle count, we assume that the initial eight random bits are available in parallel during the table lookup operation. If the random number generator is able to generate only one random bit per cycle, then additional eight cycles are required per sampling operation. However generating many (pseudo)random bits is not a problem using light-weight pseudo random number generators such as the trivium steam cipher which is used in [16]. The results in Table 4.3 show that by spending additional five slices, we can reduce the average case cycle requirement per sampling operation to almost two cycles from 17 cycles. As the sampler architecture is extremely small even with the lookup table, the acceleration provided by the fast sampling architecture will be useful in designing fast cryptosystems.

The Polynomial Sampler–1 of Table 4.3 generates a polynomial of $m = 256$ coefficients sampled from the discrete Gaussian distribution by using the fast sampler iteratively. The samples are stored in the RAM from address 0 to $m - 1$. During the consecutive sampling operations, the state-machine jumps to the next sampling operation immediately after completing a sampling operation. In this consecutive mode of sampling operations, the 'transition to the end state' cycle is not spent for the individual sampling operations. As the probability of a successful lookup operation is 0.973, in the average case 249 out of the 256 samples are generated using successful lookup operations; whereas the seven samples are obtained through the sequential bit-scanning operation. In this consecutive mode of sampling, each lookup operation generated sample consumes one cycle. Hence in the average case 249 cycles are spent for generating the majority of the samples. The seven sampling operations that perform bit scanning starting from the ninth column of the probability matrix require on average a total of 143 cycles. Thus in total 392 cycles are spent on average to generate a Gaussian distributed polynomial.

The Polynomial Sampler–2 architecture includes the random shuffling operation on a Gaussian distributed polynomial of $m = 256$ coefficients. The architecture is thus secure against simple time and power analysis attacks. However this security comes at the cost of an additional eight slices due to the requirement of additional counter and comparator circuits. The architecture first generates a polynomial in 392 cycles and then performs seven swap operations in 28 cycles in the average case. Thus in total the proposed side channel attack resistant sampler spends 420 cycles to generate a secure Gaussian distributed polynomial of 256 coefficients.

4.7 Summary

In this chapter we presented an optimized instance of the Knuth-Yao sampling archi-tecture that consumes very small area. We showed that by properly tuning the width of the ROM and the scan register, and by a decentralizing the control logic, we can reduce the area of the sampler to only 30 slices without affecting the cycle count. Moreover, we proposed a fast sampling method using a very small-area precom-puted table that reduces the cycle requirement by seven times in the average case. We showed that the basic sampler architecture can be attacked by exploiting its tim-ing and power consumption related leakages. In the end we proposed a cost-effective counter measure that performs random shuffling of the samples.

Followup works. In joint works (coauthorship) with de Clercq et al. [2] and Liu et al. [13], we adapted our approach and implemented the software versions of the Knuth-Yao algorithm on 32-bit ARM and 8-Bit AVR processors respectively. We found that the Knuth-Yao algorithm performs equally well on the software platforms.

Bruinderink et al. [7] showed that the timing leakage from a non constant-time Gaussian sampler could be exploited to break signature schemes. The work mentions that the shuffling method (Sect. 4.5.1) increases the complexity of their attack.

Pessel [15] analyzed our shuffling based countermeasure in detail and proposed a profiled side channel attack that can recover the key by observing only 7,000 signatures. He proposed to use Gaussian convolution in conjunction with shuffling to increase side channel resistance.

In a joint work with Karmakar et al. [8] (coauthorship, under review), we have proposed a constant-time implementation of the Knuth-Yao sampling algorithm. Since the Knuth-Yao random walk is dictated by a set of input random bits, we could express the sample as a function of the input random bits. Hence we represent each bit of the output sample as a Boolean expression of the random input bits. During a sampling operation these Boolean expressions are evaluated in constant-time and hence the computation time does not vary. To increase throughput, we use bit-slicing to generate multiple samples in batches.

References

1. Cormen TH, Stein C, Rivest RL, Leiserson CE (2001) Introduction to algorithms, 2nd edn. McGraw-Hill Higher Education
2. de Clercq R, Roy SS, Vercauteren F, Verbauwhede I (2015) Efficient software implementation of ring-LWE encryption. In: Proceedings of the 2015 design, automation & test in Europe conference & exhibition, DATE '15, pp 339–344
3. Devroye L (1986) Non-Uniform random variate generation. Springer, New York
4. Ducas L, Durmus A, Lepoint T, Lyubashevsky V (2013) Lattice signatures and bimodal gaussians. In: Proceedings of the 33rd annual cryptology conference advances in cryptology—CRYPTO 2013, Santa Barbara, CA, USA, 18–22 Aug 2013, Part I. Springer, Berlin, Heidelberg, pp 40–56
5. Dwarakanath N, Galbraith S (2014) Sampling from discrete gaussians for lattice-based cryptography on a constrained device. Appl Algebra Eng Commun Comput 25(3):159–180
6. Göttert N, Feller T, Schneider M, Buchmann J, Huss S (2012) On the design of hardware building blocks for modern lattice-based encryption schemes. Cryptographic hardware and embedded systems—CHES 2012. volume 7428 of LNCS. Springer, Berlin, pp 512–529
7. Groot Bruinderink L, Hülsing A., Lange T, Yarom Y (2016) Flush, gauss, and reload—a cache attack on the BLISS lattice-based signature scheme. In: Proceedings of the 18th international conference on cryptographic hardware and embedded systems—CHES 2016, Santa Barbara, CA, USA, 17–19 Aug 2016, Berlin, Heidelberg, 2016. Springer, Berlin, Heidelberg, pp 323–345
8. Karmakar A, Roy SS, Vercauteren F, Verbauwhede I (2017) Constant-time discrete gaussian sampling. Under Rev
9. Knuth DE (1997) The art of computer programming, volume 2 (3rd ed): seminumerical algorithms. Addison-Wesley Longman Publishing Co, Inc, Boston, MA, USA
10. Knuth DE, Yao AC (1976) The complexity of non-uniform random number generation. Algorithms and complexity, pp 357–428
11. Lenstra AK, Lenstra HW, Lovász L (1982) Factoring polynomials with rational coefficients. Mathematische Annalen 261(4):515–534
12. Lindner R, Peikert C (2011) Better key sizes (and Attacks) for LWE-based encryption. CT-RSA 2011:319–339
13. Liu Z, Seo H, Roy SS, Großschädl J, Kim H, Verbauwhede (2015) Efficient ring-LWE encryption on 8-bit AVR processors. In: Proceedings of the 17th international workshop on cryptographic hardware and embedded systems–CHES 2015, Saint-Malo, France, 13–16 Sept 2015, Berlin, Heidelberg. Springer, Berlin, Heidelberg, pp. 663–682
14. Lyubashevsky V (2012) Lattice signatures without trapdoors. In: Proceedings of the 31st annual international conference on theory and applications of cryptographic techniques, EURO-CRYPT'12, Berlin. Springer, pp 738–755
15. Pessl P (2016) Analyzing the shuffling side-channel countermeasure for lattice-based signatures. In: Progress in cryptology–INDOCRYPT 2016: proceeding of the 17th international conference on cryptology in India, Kolkata, India, 11–14 Dec 2016, Cham. Springer International Publishing, Cham, pp. 153–170

16. Pöppelmann T, Güneysu T (2014) Area optimization of lightweight lattice-based encryption on reconfigurable hardware. In: 2014 IEEE international symposium on circuits and systems (ISCAS), June 2014, pp 2796–2799
17. Regev O (2005) On lattices, learning with errors, random linear codes, and cryptography. In: Proceedings of the thirty-seventh annual ACM symposium on theory of computing, STOC '05, New York, NY, USA. ACM, pp 84–93

Chapter 5
Ring-LWE Public Key Encryption Processor

5.1 Introduction

In this chapter we analyze the LPR ring-LWE public key encryption scheme of Sect. 2.4.1 and design a compact hardware architecture of the encryption processor. From Fig. 2.4 of Sect. 2.4.1, we see that the LPR encryption scheme is composed of a discrete Gaussian sampler, a polynomial arithmetic (addition/multiplication) unit, a message encoder and a message decoder. In the last chapter we described how to design the discrete Gaussian sampler efficiently. In this chapter we first design a novel polynomial arithmetic unit and integrate it with the discrete Gaussian sampler to realize the ring-LWE public key encryption processor.

The polynomial arithmetic is computed in a ring $R_q = \mathbb{Z}_q[\mathbf{x}]/\langle f(x)\rangle$, where one typically chooses $f(x) = x^n + 1$ with n a power of two, and q a prime with $q \equiv 1 \bmod 2n$. An implementation thus requires the basic operations in such a ring R_q, with multiplication taking up the bulk of the resources both in area and time. An efficient polynomial multiplier architecture therefore is a pre-requisite for the deployment of ring-LWE based cryptography in real world systems. In this chapter we investigate techniques to optimize the polynomial multiplication operation in R_q.

© Springer Nature Singapore Pte Ltd. 2020   65
S. Sinha Roy and I. Verbauwhede, *Lattice-Based Public-Key Cryptography in Hardware*,
Computer Architecture and Design Methodologies,
https://doi.org/10.1007/978-981-32-9994-8_5

When we started this research, only a few hardware implementations [1, 8, 21, 23] of polynomial multipliers over R_q were known. All of these implementations have used the Number Theoretic Transform (NTT) to perform polynomial multiplication in R_q efficiently. It is well known that the Fast Fourier Transform (FFT) is asymptotically the fastest algorithm for computing polynomial multiplication [2]. NTT corresponds to a FTT where the roots of unity are taken from a finite ring instead of the complex numbers. Hence in an NTT all computations are performed on integers. The first published hardware implementation of ring-LWE encryption scheme [8] by Göttert et al. uses a fully parallel NTT structure for the polynomial multiplier resulting in a huge area consumption. For instance, even for medium security, their implementation does not fit on the largest FPGA of the Virtex 6 family. The later works [1, 21, 23] follow a sequential design methodology and use the FPGA resources in an efficient way.

In this chapter we analyze the NTT algorithm and propose several optimizations to reduce pre-computation overhead, memory requirement, and the number of memory access. We apply these optimization techniques to design a compact polynomial arithmetic core. We also perform several architectural optimizations to improve the operating frequency. Finally we connect the polynomial arithmetic core with the discrete Gaussian sampler and a memory bank to design a compact and efficient ring-LWE public key encryption processor.

The remainder of the chapter is organized as follows: In Sect. 5.2 we briefly describe the NTT algorithm and its application in computing polynomial multiplication. Section 5.3 contains our optimization techniques of the NTT and Sect. 5.4 presents the actual architecture of our optimized NTT algorithm. A pipelined architecture is given in Sect. 5.5. In Sect. 5.6, we propose an optimization of an existing ring-LWE encryption scheme and propose an efficient architecture for the complete ring-LWE encryption system. Section 5.7 reports on the experimental results of this implementation.

Target Parameter Sets

We have chosen to instantiate the cryptoprocessor for the parameter sets (n, q, s) (recall $s = \sqrt{2\pi}\sigma$), namely $P_1 = (256, 7681, 11.32)$ and $P_2 = (512, 12289, 12.18)$. Note that the choice of primes is not optimal for fast modular reduction. To estimate the security level offered by these two parameter sets we follow the security analysis in [12, 14] which improves upon [13, 28]. Apart from the dimension n, the hardness of the ring-LWE problem mainly depends on the ratio q/σ, where clearly the problem becomes easier for larger ratios. Although neither parameter set was analyzed in [14], parameter set P_1 is similar to the set $(256, 4093, 8.35)$ from [14] which requires 2^{105} seconds to break, or still over 2^{128} elementary operations. For parameter set P_2 we expect it to offer a high security level consistent with AES-256 (following [8]).

We limit the Gaussian sampler in our implementation to 12σ to obtain a negligible statistical distance ($<2^{-90}$) from the true discrete Gaussian distribution. Although one can normally sample the secret $r_2 \in R_q$ also from the distribution $\mathcal{X}_\sigma$, we restrict r_2 to have binary coefficients.

5.2 Polynomial Multiplication

Recall from Sect. 2.5.2 that NTT leads to a fast multiplication algorithm in the ring $S_q = \mathbb{Z}_q[x]/(x^n - 1)$: indeed, given two polynomials $a, b \in S_q$ we can easily compute their (reduced) product $c = a \cdot b \in S_q$ by computing

$$c = NTT_{\omega_n}^{-1}\left(NTT_{\omega_n}(a) * NTT_{\omega_n}(b)\right), \qquad (5.1)$$

where $*$ denotes point-wise multiplication.

The NTT computation is usually described as recursive, but in practice we use an in-place iterative version taken from [2] that is given in Algorithm 10. For the inverse NTT, an additional scaling of the resulting coefficients by n^{-1} is performed. The factors ω used in line 8 are called the *twiddle factors*.

Algorithm 10: *Iterative NTT*

Input: Polynomial $a(x) \in \mathbb{Z}_q[\mathbf{x}]$ of degree $n-1$ and n-th primitive root $\omega_n \in \mathbb{Z}_q$ of unity
Output: Polynomial $A(x) \in \mathbb{Z}_q[\mathbf{x}] = \text{NTT}(a)$

1 **begin**
2    $A \leftarrow BitReverse(a)$;
3   **for** $m = 2$ *to* n *by* $m = 2m$ **do**
4       $\omega_m \leftarrow \omega_n^{n/m}$;
5       $\omega \leftarrow 1$;
6   **for** $j = 0$ *to* $m/2 - 1$ **do**
7   **for** $k = 0$ *to* $n - 1$ *by* m **do**
8             $t \leftarrow \omega \cdot A[k + j + m/2]$;
9             $u \leftarrow A[k + j]$;
10             $A[k + j] \leftarrow u + t$;
11             $A[k + j + m/2] \leftarrow u - t$;
12   **end**
13          $\omega \leftarrow \omega \cdot \omega_m$;
14   **end**
15   **end**
16 **end**

Multiplication in R_q

Recall that we will use $R_q = \mathbb{Z}_q[\mathbf{x}]/\langle f \rangle$ with $f = x^n + 1$ and $n = 2^k$. Since $f(x)|x^{2n} - 1$ we could use the $2n$-point NTT to compute the multiplication in R_q at the expense of three $2n$-point NTT computations and a reduction by trivially embedding the ring R_q into S_q, i.e. expanding the coefficient vector of a polynomial $a \in R_q$ by adding n extra zero coefficients. However, we can do much better by exploiting the special relation between the roots of $x^n + 1$ and $x^{2n} - 1$ using a technique known as the *negative wrapped convolution*.

Indeed, using the same evaluation-interpolation strategy used above for the ordinary NTT, we conclude that we can efficiently multiply two polynomials $a, b \in R_q$ if we can quickly evaluate them in the roots of f. These roots are simply ω_{2n}^{2j+1} for $j = 0, \ldots, n - 1$ (since the even exponents give the roots of $x^n - 1$) and as such can be written as $\omega_{2n} \cdot \omega_n^j$. These evaluations can thus be computed efficiently using a *classical* n-point NTT (instead of a $2n$-point NTT) on the scaled polynomials $a'(x) = a(\omega_{2n} \cdot x)$ and $b'(x) = a(\omega_{2n} \cdot x)$. The point-wise multiplication gives the evaluations of $c(x) = a(x)b(x) \bmod f(x)$ in the roots of f, and the classical inverse n-point NTT thus results in the coefficients of the scaled polynomial $c'(x) = c(\omega_{2n} \cdot x)$. To recover the coefficients c_i of $c(x)$, we therefore simply have to compute $c_i = c'_i \cdot \omega_{2n}^{-i}$. Note that the scaling operation by n^{-1} can be combined with the multiplications of c'_i by ω_{2n}^{-i}.

5.3   Optimization of the NTT Computation

In this section we optimize the NTT and compare with the recent hardware implementations of polynomial multipliers [1, 21, 23]. First, the fixed cost involved in computing the powers of ω_n is reduced, then the pre-computation overhead in the forward negative-wrapped convolution is optimized, and finally an efficient memory access scheme is proposed that reduces the number of memory accesses during the NTT and also minimizes the number of block RAMs in the hardware architecture.

5.3.1   Optimizing the Fixed Computation Cost

In line 13 of Algorithm 10 the computation of the twiddle factor $\omega \leftarrow \omega \cdot \omega_m$ is performed in the j-loop. This computation can be considered as a fixed cost. However in [1, 21] the j-loop and the k-loop are interchanged, such that ω is updated in the innermost loop which is much more frequent than in Algorithm 10. To avoid the computation of the twiddle factors, in [21] all the twiddle factors are kept in a pre-computed look-up table (ROM) and are accessed whenever required. As the twiddle factors are not computed on-the-fly, the order of the two innermost loops does not result in an additional cost. However in [1] a more compact polynomial multiplier architecture is designed without using any look-up table and the twiddle factors are simply computed on-the-fly during the NTT computation. Hence in [1], the interchanged loops cause substantial additional computational overhead. In this chapter our target is to design a very compact polynomial multiplier. Hence we do not use any look-up table for the twiddle factors and follow Algorithm 10 to avoid the extra computation of [1].

5.3.2 Optimizing the Forward NTT Computation Cost

Here we revisit the forward negative-wrapped convolution technique used in [1, 21, 23]. Recall that the negative-wrapped convolution corresponds to a classical n-point NTT on the scaled polynomials $a'(x) = a(\omega_{2n} \cdot x)$ and $b'(x) = (\omega_{2n} \cdot x)$. Instead of first pre-computing these scaled polynomials and then performing a classical NTT, it suffices to note that we can integrate the scaling and the NTT computation. Indeed, it suffices to change the initialization of the twiddle factors in line 5 of Algorihtm 10: instead of initializing ω to 1, we can simply set $\omega = \omega_{2m}$. The rest of the algorithm remains exactly the same, and no pre-computation is necessary. Note that this optimization only applies to the NTT itself and not to the inverse NTT.

5.3.3 Optimizing the Memory Access Scheme

The NTT computation requires memory to store the input and intermediate coefficients. When the number of coefficients is large, RAM is most suitable for hardware implementation [1, 21, 23]. In the innermost loop (lines 8-to-11) of Algorithm 10, two coefficients $A[k + j]$ and $A[k + j + m/2]$ are first read from memory and then arithmetic operations (one multiplication, one addition and one subtraction) are performed. The new $A[k + j]$ and $A[k + j + m/2]$ are then written back to memory. During one iteration of the innermost loop, the arithmetic circuits are thus used only once, while the memory is read and written twice. This leads to idle cycles in the arithmetic circuits. The polynomial multiplier in [21] uses two parallel memory blocks to provide a continuous flow of coefficients to the arithmetic circuits. However this approach could result in under-utilization of the RAM blocks if the coefficient size is much smaller than the word size (for example in the ring-LWE cryptosystem [16]). In the literature there are many papers on efficient memory management schemes using segmentation and efficient address generation (see [17]) for the classical FFT algorithm. Another well known approach is the constant geometry FFT (or NTT) which always maintains a constant index difference between the processed coefficients [20]. However the constant geometry algorithm is not in-place and hence not suitable for resource constrained platforms. In [1] memory usage is improved by keeping two coefficients $A[k]$ and $B[k]$ of the two input polynomials A and B in the same memory location. We propose a memory access scheme which is designed to minimize the number of block RAM slices and to achieve maximum utilization of computational circuits present in the NTT architecture.

Since the two coefficients $A[k + j]$ and $A[k + j + m/2]$ are processed together in Algorithm 10, we keep the two coefficients as a pair in one memory location.

Let us analyze two consecutive iterations of the m-loop (line 3 in Algorithm 10) for $m = m_1$ and $m = m_2$ where $m_2 = 2m_1$. In the m_1-loop, for some j_1 and k_1 (maintaining the loop bounds in Algorithm 10) the coefficients $(A[k_1 + j_1], A[k_1 + j_1 + m_1/2])$ are processed as a pair. Then k increments to $k_1 + m_1$ and the processed coefficient pair is $(A[k_1 + m_1 + j_1], A[k_1 + m_1 + j_1 + m_1/2])$. Now from Algorithm 10 we see that the coefficient $A[k_1 + j_1]$ will again be processed in the m_2-loop with coefficient $A[k_1 + j_1 + m_2/2]$. Since $m_2 = 2m_1$, the coefficient $A[k_1 + j_1 + m_2/2]$ is the coefficient $A[k_1 + j_1 + m_1]$ which is updated in the m_1-loop for $k = k_1 + m_1$. Hence during the m_1-loop if we swap the updated coefficients for $k = k_1$ and $k = k_1 + m_1$ and store $(A[k_1 + j_1], A[k_1 + j_1 + m_1])$ and $(A[k_1 + j_1 + m_1/2], A[k_1 + j_1 + 3m_1/2])$ as the coefficient pairs in memory, then the coefficients in a pair have a difference of $m_2/2$ in their index and thus are ready for the m_2-loop. The operations during the two consecutive iterations $k = k_1$ and $k = k_1 + m_1$ during $m = m_1$ are shown in Algorithm 11 in lines 8–15. During the operations u_1, t_1, u_2 and t_2 are used as temporary storage registers.

Algorithm 11: *Iterative NTT: Memory Efficient Version*

Input: Polynomial $a(x) \in \mathbb{Z}_q[x]$ of degree $n - 1$ and n-th primitive root $\omega_n \in \mathbb{Z}_q$ of unity
Output: Polynomial $A(x) \in \mathbb{Z}_q[x] = \text{NTT}(a)$

1 **begin**
2      $A \leftarrow BitReverse(a)$; /* Coefficients are stored in the memory as proper pairs */
3   **for** $m = 2$ to $n/2$ by $m = 2m$ **do**
4          $\omega_m \leftarrow m$-th primitiveroot(1);
5          $\omega \leftarrow squareroot(\omega_m)$ or 1 /* Depending on forward or backward NTT */;
6   **for** $j = 0$ to $m/2 - 1$ **do**
7   **for** $k = 0$ to $n/2 - 1$ by m **do**
8                  $(t_1, u_1) \leftarrow (A[k + j + m/2], A[k + j])$ /* From MEMORY[k+j] */;
9                  $(t_2, u_2) \leftarrow (A[k + m + j + m/2], A[k + m + j])$ /* MEMORY[k+j+m/2] */;
10                 $t_1 \leftarrow \omega \cdot t_1$;
11                 $t_2 \leftarrow \omega \cdot t_2$;
12                 $(A[k + j + m/2], A[k + j]) \leftarrow (u_1 - t_1, u_1 + t_1)$;
13                 $(A[k + m + j + m/2], A[k + m + j]) \leftarrow (u_2 - t_2, u_2 + t_2)$;
14                 $MEMORY[k + j] \leftarrow (A[k + j + m], A[k + j])$;
15                 $MEMORY[k + j + m/2] \leftarrow (A[k + j + 3m/2], A[k + j + m/2])$;
16   **end**
17             $\omega \leftarrow \omega \cdot \omega_n$;
18   **end**
19   **end**
20     $m \leftarrow n$;
21     $k \leftarrow 0$;
22     $\omega \leftarrow squareroot(\omega_m)$ or 1 /* Depending on forward or backward NTT */;
23   **for** $j = 0$ to $m/2 - 1$ **do**
24         $(t_1, u_1) \leftarrow (A[j + m/2], A[j])$ /* From MEMORY[j] */;
25         $t_1 \leftarrow \omega \cdot t_1$;
26         $(A[j + m/2], A[j]) \leftarrow (u_1 - t_1, u_1 + t_1)$;
27         $MEMORY[j] \leftarrow (A[j + m/2], A[j])$;
28         $\omega \leftarrow \omega \cdot \omega_m$;
29   **end**
30 **end**

A complete description of the efficient memory access scheme is given in Algorithm 11. In this algorithm for all values of $m < n$, two coefficient pairs are processed in the innermost loop and a swap of the updated coefficients is performed before writing back to memory. For $m = n$, no swap operation is required as this is the final iteration of the m-loop. The coefficient pairs generated by Algorithm 11 can be rearranged easily for another (say inverse) NTT operation by performing address-wise bit-reverse-swap operation. Appendix A describes the memory access scheme using an example.

5.4   The NTT Processor Organization

In this section we present an architecture for performing the forward and backward NTT using the proposed optimization techniques. Our NTT processor (Fig. 5.1) consists of three main components: the arithmetic unit, the memory bloc k and the control-address unit.

The Memory Block

Is implemented as a simple dual port RAM. To accommodate two coefficients, the word size is $2\lceil \log q \rceil$ where q is the prime modulus. For the chosen parameter sets, coefficients are 13-bit or 14-bit wide. In FPGAs, a RAM can be implemented as a *distributed* or as a *block* RAM. When the amount of data is large, block RAM is the ideal choice.

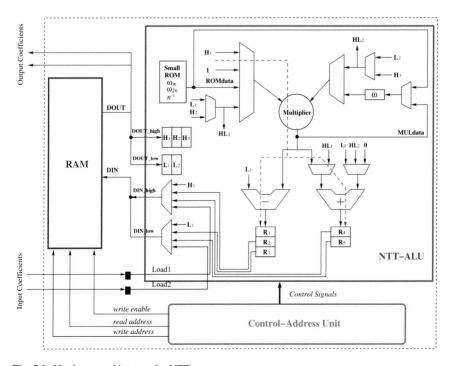

Fig. 5.1 Hardware architecture for NTT

The Arithmetic Unit (NTT-ALU)

Is designed to support Algorithm 11 along with other operations such as polynomial addition, point-wise multiplication and rearrangement of the coefficients. This NTT-ALU is interfaced with the memory block and the control-address unit. The central part of the NTT-ALU consists of a modular multiplier and addition/subtraction circuits.

Now we describe how the different components of the NTT-ALU are used during the butterfly steps (excluding the last loop for r $m = n$).

1. First, the memory location $(k + j)$ is fetched and then the fetched data (t_1, u_1) is stored in the input register pair (H_1, L_1).
2. The same also happens for the memory location $(k + j + m/2)$ in the next cycle.
3. The multiplier computes $\omega \cdot H_1$ and the result is added to or subtracted from L_1 using the adder and subtractor circuits to compute $(u_1 + \omega t_1)$ and $(u_1 - \omega t_1)$ respectively.
4. In the next cycle the register pair (R_1, R_4) is updated with $(u_1 - \omega t_1, u_1 + \omega t_1)$.
5. Another clock transition shifts the contents of (R_1, R_4) to (R_2, R_5). In this cycle the pair (R_1, R_4) is updated with $(u_2 - \omega t_2, u_2 + \omega t_2)$ as the computation involving (u_2, t_2) from the location $(k + j + m/2)$ lags by one cycle.
6. Now the memory location $(k + j)$ is updated with the register pair (R_4, R_5) containing $(u_2 + \omega t_2, u_1 + \omega t_1)$.
7. Finally, in the next cycle the memory location $(k + j + m/2)$ is updated with $(u_2 - \omega t_2, u_1 - \omega t_1)$ using the register pair (R_2, R_3).

The execution of the *last* m-loop is similar to the intermediate loops, without any data swap between the output registers. The register pair (R_2, R_5) is used for updating the memory locations. In Fig. 5.1, the additional registers $(H_2, H_3$ and $L_2)$ and multiplexers are used for supporting operations such as addition, point-wise multiplication and rearrangement of polynomials. The Small-ROM block contains the fixed values ω_m, ω_{2n}, their inverses and n^{-1}. This ROM has depth of order $\log(n)$.

The Control-and-Address Unit

consists of three counters for m, j and k in Algorithm 11 and comparators to check the terminal conditions during the execution of any loop. The read address is computed from m, j and k and then delayed using registers to generate the write address. The control-and-address unit also generates the write enable signal for the RAM and the control signals for the NTT-ALU.

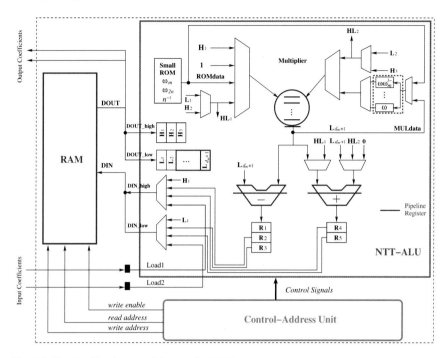

Fig. 5.2 Pipelined hardware architecture for NTT

5.5 Pipelining the NTT Processor

The maximum frequency of the NTT-ALU is determined by the critical path (red dashed line in Fig. 5.1): it passes through the modular multiplier and the adder (or subtracter) circuits. To increase the operating frequency of the processor, we implement efficient pipelines based on the following two observations.

Observation 1 During the execution of any m-loop in Algorithm 11, the computations (multiplication, addition and subtraction) involving a coefficient pair have no data dependency on other coefficient pairs. Such a data-flow structure is suitable for pipeline processing as different computations can be pipelined without inserting bubbles in the datapath.

Assume that the modular multiplier has d_m pipeline stages and that the output is latched in a buffer. In the $(d_m + 1)$th cycle after the initialization of $\omega \cdot t_1$, the buffer is updated with the result $\omega \cdot t_1$. Now we need to compute $u_1 + \omega \cdot t_1$ and $u_1 - \omega \cdot t_1$ using the adder and subtracter circuits. Hence we delay the data u_1 by d_m cycles so that it appears as an input to the adder and subtracter circuits in the $(d_m + 1)$th cycle. This delay operation is performed with the help of a shift register $L_1, \ldots, L_{d_m+1}$ as shown in Fig. 5.2.

Observation 2 Every increment of j in Algorithm 11 requires a new ω (line 17). If the multiplier has d_m pipeline stages, then the register-ω in Fig. 5.1 is updated with the new value of ω in the $(d_m + 2)$th cycle. Since this new ω is used by the next butterfly operations, the data dependency results in an interruption in the chain of butterfly operations for $d_m + 1$ cycles. In any m-loop, the total number of such *interruption cycles* is $(m/2 - 1) \cdot (d_m + 1)$.

To reduce the number of interruption cycles, we use a small look-up table to store a few twiddle factors. Let the look-up table (red dashed rectangle in Fig. 5.2) have l registers containing the twiddle factors $(\omega, \ldots \omega\omega_m^{l-1})$. This look-up table is used to provide the twiddle factors during the butterfly operations for say $j = j'$ to $j = j' + l - 1$. The next time j increments, new twiddle factors are required for the butterfly operations. We multiply the look-up table with ω_m^l to compute the next l twiddle factors $(\omega\omega_m^l, \ldots \omega\omega_m^{2l-1})$. The multiplications are independent of each other and hence can be processed in a pipeline. The butterfly operations are resumed after $\omega\omega_m^l$ is loaded in the look-up table. Thus using a small-look-up table of size l we reduce the number of interruption cycles to $(\frac{m}{2l} - 1) \cdot (d_m + 1)$. In our architecture we use $l = 4$; a larger value of l will reduce the number of interruption cycles, but will cost additional registers.

Optimal Pipeline Strategy for Speed

During the execution of any m-loop in Algorithm 11, the number of butterfly operations is $n/2$. In the pipelined NTT-ALU, the cycle requirement for the $n/2$ butterfly operations is slightly larger than $n/2$ due to an initial overhead. The state machine jumps to the ω calculation state $\frac{m}{2l} - 1$ times resulting in $(\frac{m}{2l} - 1) \cdot (d_m + 1)$ interruption cycles. Hence the total number of cycles spent in executing any m-loop can be approximated as shown below:

$$Cycles_m \approx \frac{n}{2} + \left(\frac{m}{2l} - 1\right) \cdot (d_m + 1) . \tag{5.2}$$

Assume that the delay of the critical path with no pipeline stages is D_{comb}. When the critical path is split into balanced-delay stages using pipelines, the resulting delay (D_s) can be approximated by $\frac{D_{comb}}{(d_m + d_a)}$, where d_m and d_a are the number of pipeline stages in the modular multiplier and the modular adder (subtracter) respectively. Since the delay of the modular adder is small compared to the modular multiplier, we have $d_a \ll d_m$. Now the computation time for the m-loop is approximated as

$$T_m \approx \frac{D_{comb}}{(d_m + d_a)} \left[\frac{n}{2} + \left(\frac{m}{2l} - 1\right) \cdot (d_m + 1)\right] \approx D_s \frac{n}{2} + C_m . \tag{5.3}$$

Here C_m is constant (assuming $d_a \ll d_m$) for a fixed value of m. From the above equation we find that the minimum computation time can be achieved when D_s is minimum. Hence we pipeline the datapath to achieve minimum D_s. The DSP based coefficient multiplier is optimally pipelined using the Xilinx IPCore tool, while the modular reduction block is suitably pipelined by placing registers between the cascaded adder and subtracter circuits.

5.6 The Ring-LWE Encryption Scheme

Pöppelmann et al. [23] optimized the computation cost of the LPR public key encryption scheme by keeping the fixed polynomials in the NTT domain. The message encryption and decryption operations require three and two NTT computations respectively. We reduce the number of NTT operations for decryption from two to *one*. The proposed ring-LWE encryption scheme is described below:

1. LPR.KeyGen(a): Choose a polynomial $r_1 \in R_q$ from $\mathcal{X}_\sigma$, choose another polynomial r_2 with binary coefficients and then compute $p = r_1 - a \cdot r_2 \in R_q$. The NTT is performed on the three polynomials a, p and r_2 to generate $\tilde{a}$, $\tilde{p}$ and $\tilde{r}_2$. The public key is ($\tilde{a}$, $\tilde{p}$) and the private key is $\tilde{r}_2$.
2. LPR.Encrypt($\tilde{a}$, $\tilde{p}$, m): The message m is first encoded to $\bar{m} \in R_q$. Three polynomials $e_1, e_2, e_3 \in R_q$ are sampled from $\mathcal{X}_\sigma$. The ciphertext is then computed as:

$$\tilde{e}_1 \leftarrow NTT(e_1); \quad \tilde{e}_2 \leftarrow NTT(e_2)$$
$$(\tilde{c}_1, \tilde{c}_2) \leftarrow \left(\tilde{a} * \tilde{e}_1 + \tilde{e}_2; \ \tilde{p} * \tilde{e}_1 + NTT(e_3 + \bar{m})\right).$$

3. LPR.Deccrypt($\tilde{c}_1, \tilde{c}_2, \tilde{r}_2$) : Compute m' as $m' = INTT(\tilde{c}_1 * \tilde{r}_2 + \tilde{c}_2) \in R_q$ and recover the original message m from m' using a decoder.

The scheme requires both encryption and decryption to use a common primitive root of unity.

5.6.1 Hardware Architecture

Figure 5.3 shows the hardware architecture for the ring-LWE encryption system. The basic building blocks used in the architecture are: the memory file, the arithmetic unit, the discrete Gaussian sampler and the control-address generation unit. The arithmetic unit is the NTT-ALU that we described in the previous section. Here we briefly describe the memory file and the discrete Gaussian sampler.

The Memory File is designed to support the maximum memory requirement that occurs during the encryption of the message. Six memory blocks M_0 to M_5 are available in the memory file and are used to store $\bar{a}$, $\bar{p}$, e_1, e_2, e_3 and $\bar{m}$ respectively. The memory blocks have width $2\lceil \log q \rceil$ bits and depth $n/2$. All six memory blocks share a common read and a write address and have a common data-input line, while their data-outputs are selected through a multiplexer. Any of the memory blocks in the memory file can be chosen for read and write operation. Due to the common addressing of the memory blocks, the memory file supports one read and one write operation in every cycle.

The Discrete Gaussian Sampler is based on the Knuth-Yao sampler architecture that we designed in the last chapter. The sampler does not include the shuffling coun-

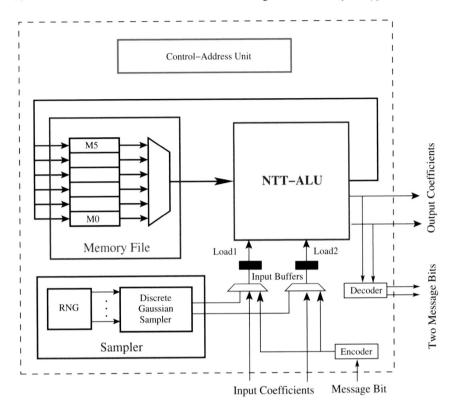

Fig. 5.3 Ring-LWE cryptoprocessor

termeasure as in this work we intend to design the core of the public key encryption processor and measure its performance without the overhead of countermeasures. The sampler architecture has a sufficiently large precision and tail-bound to satisfy a maximum statistical distance of 2^{-90} to a true discrete Gaussian distribution for both $s = 11.32$ and $s = 12.18$. Two look-up tables are used to speedup the sampling operation. The first lookup table maps eight random bits and the second lookup table maps five random bits. When the second lookup operation fails (probability <0.0016) then bit-scan based Knuth-Yao random walk is started with the initial distance obtained from the second lookup operation.

The Cycle Count for the encryption and decryption operations can be minimized in the following way. During the encryption operation, first the three error polynomials e_1, e_2 and e_3 are generated by invoking the discrete Gaussian sampler $3n$ times. Next the encoded message $\bar{m}$ is added to e_3 and then three consecutive forward NTT operations are performed on e_1, e_2 and $(e_3 + \bar{m})$. Finally the ciphertext $\tilde{c}_1, \tilde{c}_2$ is obtained using two coefficient-wise multiplications followed by two polynomial additions and two rearrangement operations. The decryption operation requires one coefficient-wise multiplication, one polynomial addition and finally one inverse NTT operation.

During the encryption operation, $3n$ samples are generated to construct the three error polynomials. Our fast Knuth-Yao sampler architecture requires 805 and 1644 cycles for the dimensions 256 and 512 respectively on average to generate the three error polynomials. The polynomial addition and point-wise multiplication operations require n cycles each with a small overhead. The consecutive processing of I forward NTTs share a fixed computation cost fc_{fwd} and require in total $fc_{fwd} + I \times \frac{n}{2} \log(n)$ cycles. Similarly I consecutive inverse NTTs are processed in $fc_{inv} + I \times \frac{n}{2} \log(n) + I \times n$ cycles. One interesting point is that the fixed cost fc_{inv} is larger than fc_{fwd} as it includes the computation of ω_{2n}^i/N (Sect. 5.2) for $i = (0 \ldots n-1)$. This observation has been used to optimize the overall ring-LWE based encryption scheme in Sect. 5.6. The additional $I \times n$ cycles during the inverse NTTs are required to multiply the coefficients by the scaling factors. The rearrangement of polynomial coefficients after an NTT operation requires less than n cycles. From the above cycle counts for each primitive operations, we see that the encryption and decryption operations require total $fc_{fwd} + \frac{3}{2}n \log(n) + 10n$ and $fc_{inv} + \frac{n}{2} \log(n) + 3n$ cycles respectively along with additional overhead. Our ring-LWE architecture has the fixed computation costs $fc_{fwd} = 667$ and $fc_{inv} = 1048$ cycles for $n = 256$; and $fc_{fwd} = 1139$ and $fc_{inv} = 1959$ cycles for $n = 512$.

5.7 Experimental Results

We have implemented the LPR ring-LWE cryptosystem on the Xilinx Virtex 6 FPGA for the parameter sets (n, q, s): $(256, 7681, 11.32)$ and $(512, 12289, 12.18)$. The area and performance results are obtained from the Xilinx ISE12.2 tool after place and route analysis and are shown in Table 5.1. In the table we also compare our results with other reported hardware implementations of the ring-LWE encryption scheme.

Our implementations are both fast and small due to the proposed computational optimizations and resource efficient design style. The cycle counts shown in the table do not include the cycles for data loading or reading operations. Our Knuth-Yao samplers have less than 2^{-90} statistical distances from the corresponding true discrete Gaussian distributions and consume around 164 LUTs and have delay less than 2.5 ns (with optimization goal for speed). Such a small delay makes the sampler suitable for integration in the pipelined ring-LWE processor under a single clock domain. We use nine parallel true random bit generators [6, 7] to generate the random bits for the sampler. The set of true random bit generators consumes 378 LUTs and 9 FFs.

The first hardware implementation of the LPR ring-LWE encryption scheme in [8] uses a heavily parallel architecture to minimize the number of clock cycles for the NTT computation. Due to the many parallel computational blocks, the architecture is very large (0.29 million LUTs and 0.14 million FFs for $n = 256$) and does not even fit on the largest FPGA of the Virtex 6 family. Performance results such as cycle count and frequency are not reported in their paper. The architecture uses a Gaussian distributed array for sampling of the error coefficients up to a tail-bound of $\pm 2s$.

The implementation in [23] is small and fast due to its resource-efficient design style. A high operating frequency is achieved using pipelines in the architecture. The architecture uses a ROM that keeps all the twiddle factors required during the NTT operation. This approach reduces the fixed computation cost (fc) but consumes block RAM slices in FPGAs. Additionally, the parallel RAM blocks in the NTT processor result in a larger memory requirement compared to our design. The discrete Gaussian sampler is based on the inversion sampling method [5] and has a maximum statistical distance of 2^{-22} to a true discrete Gaussian distribution. Since the inversion sampling requires many random bits to output a sample value, an AES core is used as a pseudo-random number generator. The AES core itself consumes an additional 803 LUTs and 341 FFs compared to our true random number generator. Another reason behind the larger area consumption of [23] compared to our architecture is due to the fact that the architecture supports different parameter sets at synthesis time. Our ring-LWE processor is also designed to achieve scalability for various parameter sets. In our architecture the control block remains the same; while only the data-width and the modular reduction block changes for different parameter sets. Hence our architecture is also configurable by generating the HDL codes for various parameter sets using a C program.

Although our architecture does not use a dedicated ROM for storing the twiddle factors, it still achieves slightly smaller cycle count and faster computation time compared to [23]. The encryption scheme in [23] computes one forward and two inverse NTTs; while our encryption scheme computes only forward NTTs and hence does not require the $4n$ cycles for the scaling operation. Additionally our negative convolution method is free from the precomputation that takes n cycles in [23]. Hence we save $5n$ cycles in total during the NTT operations in an encryption operation. Since

Table 5.1 Performance and comparison

Implementation algorithm	Parameters	Device	LUTs/FFs/DSPs/ BRAM18	Freq (MHz)	Cycles/Time (μs)	
					Encryption	Decryption
Our RLWE	(256, 7681, 11.32)	V6LX75T	1349/860/1/2	313	6.3 k/20.1	2.8 k/9.1
Our RLWE	(512, 12289, 12.18)		1536/953/1/3	278	13.3 k/47.9	5.8 k/21
RLWE [23]	(256, 7681, 11.32)	V6LX75T	4549/3624/1/12	262	6.8 k/26.2	4.4 k/16.8
RLWE	(512, 12289, 12.18)	V6LX75T	5595/4760/1/14	251	13.7 k/54.8	8.8 k/35.4
RLWE-Enc [22] RLWE-Dec	(256, 4096,8.35)	S6LX9	317/238/95/1 112/87/32/1	144 189	136 k/946 –	– 66k/351
ECC [24]	Binary-233	V5LX85T	18097/-/5644/0	156	1.9 k/12.3	1.9 k/12.3
NTRU [11]	NTRU-251	XCV1600E	27292/5160/14352/0	62.3	–/1.54	–/1.41

the fixed computation cost fc_{fwd} is smaller than $5n$, we gain in cycle count for the encryption operation. The decryption operation in our case is trivially faster than [23] as only one NTT is performed. We also reduce the area and memory requirement significantly compared to [8, 23]. This reduction is achieved by our resource-efficient design decisions such as (1) absence of a dedicated ROM for the twiddle factors, (2) an efficient RAM access and storage scheme, (3) use of one modular multiplier, (4) use of a smaller and faster (low-delay) discrete Gaussian sampler, and finally (5) the resource sharing between different computations.

The lightweight implementation [22] proposes ring-LWE encryption and decryption architectures targeting small area at the cost of performance. The implementation uses a quadratic-complexity multiplier instead of a complicated NTT based polynomial multiplier. Additionally the special modulus also saves some amount of area as the modular reduction is free of cost. However if we consider a similar quadratic-complexity multiplication based architecture in the dimension $n = 512$, then the cycle requirement will be nearly 40 times compared to our NTT-based ring-LWE processor. Our target was to use FPGA resources more efficiently without affecting the performance and to achieve similar speed as [23].

We also compare our results with other cryptosystems such as ECC and NTRU. The ECC processor [24] over the NIST recommended binary field $GF(2^{233})$ requires 12.3 μs to compute one scalar multiplication and is faster than our ring-LWE processor. However the ECC processor is designed to achieve high speed and hence consumes very large area compared to our ring-LWE processor. The NTRU scheme [11] is much faster than our ring-LWE processor due to its less complicated arithmetic. However the parameters chosen for the implementation in [11] have security around 64 bits [10]. Though secure parameter sets for the NTRU based encryption have been proposed in [9], no hardware implementation for the secure parameter sets is available in the literature.

5.8 Summary

In this chapter we analyzed the NTT based polynomial multiplication algorithm and proposed several optimizations to increase its computational efficiency and reduce storage requirement. We applied these optimization tricks to design a compact hardware architecture for polynomial arithmetic in the ring-LWE encryption scheme. We finally integrated the polynomial arithmetic unit with the compact Knuth-Yao discrete sampler from the last chapter and designed a compact and efficient ring-LWE public key encryption processor.

The design methodology and the optimizations make the cryptoprocessor architecture suitable for resource-constrained platforms. Although the chapter focuses on implementation of the ring-LWE based encryption system, we finally remark that the proposed optimization techniques for the NTT computation are applicable for other lattice based cryptosystems where similar polynomial multiplications are performed. In the next chapters, we will design processors for homomorphic encryption schemes. There we will show that the proposed optimizations could be very helpful in reducing the computation time.

Followup works In joint works (coauthorship) with de Clercq et al. [3] and Liu et al. [15], we implemented the LPR ring-LWE public key encryption on 32-bit ARM and 8-Bit AVR processors respectively. The proposed optimizations in this chapter were adapted in [3] for the 32-bit processor architecture. For the parameter set $(n, q, s) = (256, 7681, 11.32)$ the software requires 121,166 cycles per encryption and 43,324 cycles per decryption. The encryption would be an order of magnitude faster than an implementation of the elliptic curve based ECIES encryption scheme (see Sect. 2.2) on a similar platform if the elliptic curve point multiplier of [4] is used. The software by Liu et al. shows that fast implementation of the ring-LWE encryption scheme is feasible on resource-constrained 8-bit AVR processors.

Side channel analysis of a new cryptographic constructions has always received interest from the research community. In joint works (coauthorship) with Reparaz et al. [25–27] we developed masking based countermeasures against differential power analysis attacks. Oder et al. [18] proposed practical ring-LWE based public key encryption that is protected against adaptive chosen-ciphertext attacks and equipped with countermeasures against side channel attacks. Park et al. [19] mounted an SPA attack combined with chosen ciphertext attack on the ring-LWE encryption. The attack exploits the computational variations during modular additions on 8-bit processors.

References

1. Aysu A, Patterson C, Schaumont P (2013) Low-cost and area-efficient fpga implementations of lattice-based cryptography. In: 2013 IEEE international symposium on hardware-oriented security and trust (HOST), pp 81–86
2. Cormen TH, Stein C, Rivest RL, Leiserson CE (2001) Introduction to algorithms, 2nd edn. McGraw-Hill Higher Education, Pennsylvania
3. de Clercq R, Roy SS, Vercauteren F, Verbauwhede I (2015) Efficient software implementation of ring-LWE encryption. In: Proceedings of the 2015 design, automation & test in europe conference & exhibition, DATE '15, pp 339–344
4. de Clercq R, Uhsadel L, Van Herrewege A, Verbauwhede I (2014) Ultra Low-Power Implementation of ECC on the ARM Cortex-M0+. In: Proceedings of the 51st annual design automation conference, DAC '14. ACM, New York, NY, USA, pp 112:1–112:6
5. Devroye L (1986) Non-uniform random variate generation. Springer, New York
6. Dichtl M, Golic JD (2007) High-speed true random number generation with logic gates only. Cryptographic hardware and embedded systems - CHES 2007, LNCS, vol 4727. Springer, Berlin, pp 45–62
7. Golic JD (2006) New methods for digital generation and postprocessing of random data. IEEE Trans Comput 55(10):1217–1229
8. Göttert N, Feller T, Schneider M, Buchmann J, Huss S (2012) On the design of hardware building blocks for modern lattice-based encryption schemes. Cryptographic hardware and embedded systems-CHES 2012. LNCS, vol 7428. Springer, Berlin, pp 512–529
9. Hirschhorn P, Hoffstein J, Howgrave-graham N, Whyte W (2009) Choosing NTRUEncrypt parameters in light of combined lattice reduction and MITM approaches. In: Proceedings of the ACNS 2009, LNCS, vol 5536, Springer, pp 437–455

10. Howgrave Graham N (2007) A hybrid lattice-reduction and meet-in-the-middle attack against NTRU. In: Advances in cryptology - CRYPTO 2007. Lecture notes in computer science, vol 4622. Springer, Berlin, pp 150–169
11. Kamal A, Youssef A (2009) An FPGA implementation of the NTRUEncrypt cryptosystem. In: 2009 international conference on microelectronics (ICM), pp 209–212
12. Lepoint T, Naehrig M (2014) A comparison of the homomorphic encryption schemes FV and YASHE. In: Progress in cryptology – AFRICACRYPT 2014: 7th international conference on cryptology in Africa, Marrakesh, Morocco, 28–30 May 2014. Springer International Publishing, Cham, pp 318–335
13. Lindner R, Peikert C (2011) Better key sizes (and Attacks) for LWE-based encryption. CT-RSA 2011:319–339
14. Liu M, Nguyen PQ (2013) Solving BDD by enumeration: an update. In: Proceedings of the 13th international conference on topics in cryptology, CT-RSA'13. Springer, Berlin, pp 293–309
15. Liu Z, Seo H, Roy SS, Großschädl J, Kim H, Verbauwhede I (2015) Efficient ring-LWE encryption on 8-Bit AVR processors. In: Proceedings of the 17th international workshop cryptographic hardware and embedded systems – CHES 2015, Saint-Malo, France, 13–16 September 2015. Springer, Berlin, pp 663–682
16. Lyubashevsky V, Peikert C, Regev O (2010) On ideal lattices and learning with errors over rings. Advances in cryptology-EUROCRYPT 2010. Lecture notes in computer science, vol 6110. Springer, Berlin, pp 1–23
17. Ma Y, Wanhammar L (2000) A hardware efficient control of memory addressing for high-performance FFT processors. IEEE Trans Signal Process 48(3):917–921
18. Oder T, Schneider T, Pöppelmann T, Güneysu T (2015) Practical CCA2-secure and masked ring-LWE implementation. Cryptology ePrint Archive, Report 2016/1109. http://eprint.iacr.org/2016/1109
19. Park A, Han DG (2016) chosen ciphertext simple power analysis on software 8-bit implementation of ring-LWE encryption. In: 2016 IEEE Asian hardware-oriented security and trust (AsianHOST), pp 1–6
20. Pollard J (1971) The fast fourier transform in a Finite Field. Math Comput 25:365–374
21. Pöppelmann T, Güneysu T (2012) Towards efficient arithmetic for lattice-based cryptography on reconfigurable hardware. Progress in cryptology-LATINCRYPT 2012. LNCS, Vol 7533. Springer, Berlin, pp 139–158
22. Pöppelmann T, Güneysu T (2014) Area optimization of lightweight lattice-based encryption on reconfigurable hardware. In: 2014 IEEE international symposium on circuits and systems (ISCAS), pp 2796–2799
23. Pöppelmann T, Güneysu T (2014) Towards practical lattice-based public-key encryption on reconfigurable hardware. Selected areas in cryptography - SAC 2013. Lecture notes in computer science. Springer, Berlin, pp 68–85
24. Rebeiro C, Roy SS, Mukhopadhyay D (2012) Pushing the limits of high-speed $GF(2^m)$ elliptic curve scalar multiplication on FPGAs. In: Proceedings of the 14th international workshop cryptographic hardware and embedded systems – CHES 2012, Leuven, Belgium, 9–12 Sept 2012. Springer, Berlin, pp 494–511
25. Reparaz O, de Clercq R, Roy SS, Vercauteren F, Verbauwhede I (2016) Additively homomorphic ring-LWE masking. In: Post-quantum cryptography: 7th international workshop, PQCrypto 2016, Fukuoka, Japan, 24–26 Feb 2016, Proceedings. Springer International Publishing, Cham, pp 233–244
26. Reparaz O, Roy SS, de Clercq R, Vercauteren F, Verbauwhede I (2016) Masking ring-LWE. J Cryptograph Eng 6(2):139–153
27. Reparaz O, Roy SS, Vercauteren F, Verbauwhede I (2015) A masked ring-LWE implementation. In: Cryptographic hardware and embedded systems – CHES 2015: 17th international workshop, Saint-Malo, France, proceedings, 13–16 Sept 2015. Springer Berlin, pp 683–702
28. van de Pol J, Smart NP (2013) Estimating key sizes for high dimensional lattice-based systems. In: IMA international conference. Lecture notes in computer science, vol 8308. Springer, pp 290–303

Chapter 6
Conclusions and Future Work

In this chapter we summarize the contributions of this work and point out some of the possible future directions.

6.1 Conclusions

Strong ECC is feasible on IoT Efficient implementations of elliptic-curve cryptography (ECC) targeting different application requirements have received interest for over two decades. However the proposals for implementing ECC on tiny devices focused predominately on 163 bit elliptic-curves which provide only 80 bit security. Feasibility of larger elliptic-curves on such devices was not investigated. In this research we designed a 140 bit secure lightweight ECC architecture based on a 283 bit Koblitz curve with countermeasures against timing and power side channel attacks. When instantiated as a coprocessor of commercial 16 bit microcontrollers, the ECC architecture consumes only 4.3 KGE, showing its potential use in low-end Internet of things (IoT) devices.

Ring-LWE-based PKC is fast Construction of public-key cryptography (PKC) primitives that are secure against quantum computers is a very recent topic. In this research we investigated the implementation aspects of public-key encryption based on the *ring learning with errors* (ring-LWE) problem, which is presumed to be secure against quantum computers. We analyzed the arithmetic primitives, namely discrete Gaussian sampling and polynomial arithmetic. We showed that high precision discrete Gaussian sampling can be implemented in hardware using a very small amount of resources following an adaptation of the Knuth–Yao algorithm. Our sampler architecture is also very fast. For polynomial multiplication, we used the NTT method coupled with additional optimizations in the computation steps and the architecture. As a result of our design decisions and optimization strategies, the implemented public-key encryption processor achieves very fast computation time (48/21 μs per encryption/decryption) while using minimum area and memory.

© Springer Nature Singapore Pte Ltd. 2020
S. Sinha Roy and I. Verbauwhede, *Lattice-Based Public-Key Cryptography in Hardware*,
Computer Architecture and Design Methodologies,
https://doi.org/10.1007/978-981-32-9994-8_6

6.2   Future Works

ECC-ring-LWE hybrid schemes There are new publications [2, 3] that propose hybrid public-key schemes for present-day applications: ring-LWE for public-key encryption or key exchange and ECC for digital signature. The idea is that encryption or key exchange schemes should remain secure against future quantum computing attacks; whereas for digital signature based authentication schemes, security against the existing cryptanalysis techniques is sufficient. Hence it would be interesting to design a unified cryptoprocessor by combining the Koblitz curve processor of Chap. 3 with with the ring-LWE-based public-key encryption processor of Chap. 5.

Post-quantum digital signature schemes In this research we investigated the implementation aspects of ring-LWE-based public-key and homomorphic encryption schemes. There are several post-quantum digital signature schemes that work on polynomial rings. The main challenge in some of these schemes is that they require sampling from a discrete Gaussian distribution with large standard deviation. Pöppelmann, Ducas and Güneysu [4] showed how to efficiently sample from such wide distributions. Still it will be interesting to develop more efficient sampling algorithms that achieve better performance. New signature schemes, such as TESLA [1], do not require Gaussian sampling during signature generation. It will be interesting to evaluate their performance on hardware platforms.

Post-quantum cryptography for IoT From this research we can conclude that ring-LWE problem based public-key cryptography is as computationally intensive as the classical schemes. IoT devices are constrained by the amount of available resources such as computation capability, storage or memory, power and energy consumption.

In this research, we mainly investigated the implementation aspects of ring-LWE-based post-quantum public-key cryptography and performed implementation specific optimizations targeting fast computation time. An interesting direction for future work would be to investigate lightweight design methodology taking into account the limitations of IoT devices such as small silicon area and low energy/power consumption. This would require mathematical or system-level optimizations and modifications tailored towards IoT. Ring-LWE-based cryptography relies a lot on polynomial arithmetic. Hence it would be interesting to design algorithms that could perform polynomial arithmetic by consuming a very small amount of resources.

Protection against physical attacks In this research we developed efficient algorithms and architectures for ring-LWE-based cryptographic schemes. Designing of countermeasures against side channel and fault attacks would be a very interesting future research.

Effect of bias in randomness on Gaussian sampling The discrete Gaussian sampler requires random numbers. Any bias in the randomness would result in a large statistical distance to the accurate Gaussian distribution, and this could be exploited by an attacker. Hence it would be interesting to study the effect of a biased random number generator on the Gaussian sampling.

References

1. Alkim E, Bindel N, Buchmann J, Dagdelen ö, Schwabe P (2015) TESLA: tightly-secure efficient signatures from standard lattices. Cryptology ePrint archive, report 2015/755. http://eprint.iacr.org/2015/755
2. Bos JW, Costello C, Naehrig M, Stebila D (2015) Post-quantum key exchange for the TLS protocol from the ring learning with errors problem. In: 2015 IEEE symposium on security and privacy, pp 553–570
3. Bos J, Costello C, Ducas L, Mironov I, Naehrig M, Nikolaenko V, Raghunathan A, Stebila D (2016) Frodo: take off the ring! practical, quantum-secure key exchange from LWE. In: Proceedings of the 2016 ACM SIGSAC conference on computer and communications security, CCS '16, New York, NY, USA, pp 1006–1018. ACM
4. Pöppelmann T, Ducas L, Güneysu T (2014) Enhanced lattice-based signatures on reconfigurable hardware. In: Cryptographic hardware and embedded systems–CHES 2014: 16th international workshop. Busan, South Korea, September 23–26, pp 353–370. Proceedings. Springer, Berlin, Heidelberg

Appendix A
High Speed Scalar Conversion for Koblitz Curves

Here we show that optimization tricks similar to Sect. 3.2 can be applied to design a high speed scalar conversion algorithm. We choose the high-speed variant of the lazy reduction algorithm [1] known as the *double lazy reduction*. The algorithm is based on the fact that division by τ^2 is also easy to perform in hardware following Theorem 2 in Sect. 2.2.1. During a scalar reduction, the algorithm performs repeated divisions by τ^2 for $(m - 1)/2$ number of times. As a result, the cycle requirement reduces to nearly half compared to the lazy reduction. The computational steps performed in the double lazy reduction are shown in Algorithm 12.

Elimination of Long Subtractions for Nonzero Remainders

In line 6 of Algorithm 12, remainders u_0 and $u_1 \in \{0, 1\}$ are subtracted from d_0 and d_1. We observe that the subtractions are *easy* in some cases. For example, when $d_0 \equiv 1 \pmod 4$ and $2d_1 \equiv 0 \pmod 4$ (i.e. $u_0 = 1$ and $u_1 = 0$), the subtraction of u_0 from d_0 is equivalent to changing the least significant bit of d_0 from 1 to 0. Hence, in this case the long subtraction can be replaced by a bit alteration. However, when carry propagations are involved with long subtractions, alteration of few specific bits do not work as a replacement. For example, when $d_0 \equiv 3 \pmod 4$ and $2d_1 \equiv 0 \pmod 4$ (i.e. when $u_0 = 1$ and $u_1 = 1$), a long subtraction appears. Use of signed remainder set u_0 and $u_1 \in \{0, \pm 1\}$ helps to a certain extent in eliminating the long subtractions of nonzero remainders for such cases. Table A.1 shows how the signed remainders are generated during the reduction steps depending on the low bits of d_0 and $2d_1$. It can be noticed from the table that, except Case 4, subtractions of the u_0 and u_1 from d_0 and d_1 involve no carry propagation and thus can be replaced by alterations of the low bits in d_0 and d_1.

For Case 4, if we perform the subtraction of $u_0 = -1$ in line 7 of Algorithm 12 instead of line 6 (i.e. we put $d_0 + 1$ in place of d_0), then we have the following observation.

© Springer Nature Singapore Pte Ltd. 2020
S. Sinha Roy and I. Verbauwhede, *Lattice-Based Public-Key Cryptography in Hardware*,
Computer Architecture and Design Methodologies,
https://doi.org/10.1007/978-981-32-9994-8

Algorithm 12: *Fast scalar reduction from [1]*

Input: integer k
Output: reduced scalar γ
1 **begin**
2      $(a_0, a_1) \leftarrow (1, 0),\ (b_0, b_1) \leftarrow (0, 0),\ (d_0, d_1) \leftarrow (k, 0)$;
3   **for** $i = 1$ *to* $(m - 1)/2$ **do**
4         $u \leftarrow (d_0 - 2d_1) \bmod 4$;
5         $u_0 \leftarrow u \bmod 2,\ u_1 \leftarrow \lfloor u/2 \rfloor$;
6         $d_0 \leftarrow d_0 - u_0,\ d_1 \leftarrow d_1 - u_1$;
7         $(d_0, d_1) \leftarrow ((-d_0 - 2d_1)/4, -(-d_0 + 2d_1)/4)$;
8   **if** $u > 0$ **then**
9            $b_0 \leftarrow (b_0 + u_0 a_0 - 2u_1 a_1)$;
10           $b_1 \leftarrow b_1 + u_0 a_1 + u_1(a_0 - a_1)$;
11   **end**
12        $(a_0, a_1) \leftarrow (-2(a_0 - a_1),\ -a_0 - a_1)$;
13   **end**
14   **if** $d_0 \equiv 1 (\bmod\ 2)$ **then**
15        $u \leftarrow 1,\ d_0 \leftarrow d_0 - 1$;
16        $(b_0, b_1) \leftarrow (b_0 + a_0, b_1 + a_1)$;
17   **end**
18     $(d_0, d_1) \leftarrow (-d_0/2 + d_1, -d_0/2)$;
19     $\gamma \leftarrow (b_0 + d_0, b_1 + d_1)$;
20 **end**

Table A.1 Signed remainders during reduction of scalar

Cases	$d_0\ (mod\ 4)$	$2d_1\ (mod\ 4)$	u_0	u_1
1	0	0	0	0
2	1	0	1	0
3	2	0	0	−1
4	3	0	−1	0
5	0	2	0	1
6	1	2	−1	0
7	2	2	0	0
8	3	2	1	0

$$d_0 \leftarrow -\frac{2d_1 + (d_0 + 1)}{4}$$

$$d_1 \leftarrow -\frac{2d_1 - (d_0 + 1)}{4}$$

This is equivalent to taking carry/borrow inputs in the adder/subtracter circuits during the computations of d_0 and d_1. This is shown below.

$$d_0 \leftarrow -\frac{2d_1 + d_0 + (Carry\ input\ =\ 1)}{4}$$

$$d_1 \leftarrow -\frac{2d_1 - d_0 - (Borrow\ input\ =\ 1)}{4}$$

From the observations presented in this subsection, we draw the conclusion that the long subtractions of the nonzero remainders can be eliminated by changing the low order bits of d_0 and d_1 or by considering carry/borrow inputs to the adder/subtracter circuits.

Elimination of Subtractions from Zero

In line 7 of Algorithm 12, a subtraction from zero is required for d_0 after computing $2d_1 + d_0$ (Eq. A.1).

$$(d_0, d_1) \leftarrow \left(-\frac{2d_1 + d_0}{4}, -\frac{2d_1 - d_0}{4} \right) \tag{A.1}$$

We eliminate the subtraction from zero using the following scheme. Instead of Eq. A.1, we compute Eq. A.2.

$$(d_0, d_1) \leftarrow \left(\frac{2d_1 + d_0}{4}, \frac{2d_1 - d_0}{4} \right) \tag{A.2}$$

The results from Eqs. A.1 and A.2 have opposite signs, but same magnitudes. So, when Eqs. A.1 and A.2 are computed inside the for-loop in Algorithm 12 for an even number of times, the results of the equations are same; otherwise the results have opposite signs.

The same trick is applied to eliminate the subtractions from zero during the computation of (a_0, a_1) in line 12 of Algorithm 12. Instead of computing Eq. A.3

$$(a_0, a_1) \leftarrow (-2(a_0 - a_1), -(a_0 + a_1)) \tag{A.3}$$

we compute Eq. A.4.

$$(a_0, a_1) \leftarrow (2(a_0 - a_1), (a_0 + a_1)) \tag{A.4}$$

After completion of the for-loop, one subtraction from zero is required to make the signs correct for (a_0, a_1) when $(m - 1)/2$ is odd. This subtraction from zero can be eliminated if we compute Eq. A.5 in line 16 of Algorithm 12.

$$(b_0, b_1) \leftarrow (b_0 - a_0, b_1 - a_1) \tag{A.5}$$

The improved high-speed scalar reduction algorithm is described in Algorithm 13.

Algorithm 13: *New Reduction Algorithm*

Input: integer k
Output: reduced scalar γ

1 **begin**
2 | $(a_0, a_1) \leftarrow (1, 0)$, $(b_0, b_1) \leftarrow (0, 0)$, $(d_0, d_1) \leftarrow (k, 0)$;
3 | /* Iterative divisions by τ^2 start here */ ;
4 | **for** $i = 1$ *to* $(m - 1)/2$ **do**
5 | | $u \leftarrow (d_0 - 2d_1) \bmod 4$;
6 | | $(u_0, u_1) \leftarrow Table\ 1$;
7 | | $(d_0, d_1) \leftarrow Alter\,Low\,Bits(d_0, d_1)$;
8 | | **if** *Case 4 is True* **then**
9 | | | $(B, C) \leftarrow (1, 1)$ /* Borrow and Carry Inputs */
10 | | **end**
11 | | **else**
12 | | | $(B, C) \leftarrow (0, 0)$
13 | | **end**
14 | | $(d_0, d_1) \leftarrow ((2d_1 + d_0 + C)/4, (2d_1 - d_0 - B)/4)$;
15 | | **if** $u > 0$ **then**
16 | | | $b_0 \leftarrow (b_0 + u_0 a_0 - 2u_1 a_1)$;
17 | | | $b_1 \leftarrow (b_1 + u_0 a_1 + u_1(a_0 - a_1))$;
18 | | **end**
19 | | $(a_0, a_1) \leftarrow (2(a_0 - a_1),\ a_0 + a_1)$;
20 | **end**
21 | /* Iterative divisions by τ^2 finish here */ ;
22 | **if** $d_0 \equiv 1(mod\ 2)$ **then**
23 | | $d_0 \leftarrow Alter\,Least\,Bit(d_0)$;
24 | | **if** $\frac{m-1}{2} \equiv 0(mod\ 2)$ **then**
25 | | | $(b_0, b_1) \leftarrow (b_0 + a_0, b_1 + a_1)$;
26 | | **end**
27 | | **else**
28 | | | $(b_0, b_1) \leftarrow (b_0 - a_0, b_1 - a_1)$;
29 | | **end**
30 | **end**
31 | $(d_0, d_1) \leftarrow ((2d_1 - d_0)/2, d_0/2)$ /* Final division by τ */ ;
32 | **if** $\frac{m-1}{2} \equiv 0(mod\ 2)$ **then**
33 | | $\gamma \leftarrow (b_0 + d_0, b_1 - d_1)$;
34 | **end**
35 | **else**
36 | | $\gamma \leftarrow (b_0 - d_0, b_1 + d_1)$;
37 | **end**
38 **end**

A.1 Improved Double Digit τNAF Generation

In [1], two consecutive τNAF digits are generated in a single step from the reduced scalar $d_0 + \tau d_1$ by performing divisions by τ^2. The authors call the NAF as *double* τNAF. Table A.2 shows how the consecutive τNAF digits r_0 and r_1 are generated by observing the low order bits of d_0 and d_1. Similar to Sect. 3.2.1, we eliminate the subtractions of nonzero remainders from d_0 and d_1 during the τNAF generation process.

From Table A.2, we see that for the cases 2, 3.B, 3.C, 3.D, 5.A, 5.B, 5.D, 6 and 8, the subtractions of nonzero remainders from d_0 or d_1 affect only the low order bits of d_0 and d_1. For the above cases, the long subtractions are replaced by cheaper

Table A.2 NAF generation for $\mu = -1$

Cases	$d_0 (mod 4)$	$2d_1 (mod 4)$	$d_0 (mod 8)$	$2d_1 (mod 8)$	r_0	r_1
1	0	0			0	0
2	1	0			1	0
3.A	2	0	2	0	0	1
3.B	2	0	6	0	0	−1
3.C	2	0	2	4	0	−1
3.D	2	0	6	4	0	1
4	3	0			−1	0
5.A	0	2	0	2	0	1
5.B	0	2	4	2	0	−1
5.C	0	2	0	6	0	−1
5.D	0	2	4	6	0	1
6	1	2			−1	0
7	2	2			0	0
8	3	2			1	0

bit alterations in d_0 and d_1. Subtraction of $r_0 = -1$ in Case 4 in Table A.2 can be handled in the same way we did for Case 4 in Table A.1 (Sect. 3.2).

In Case 3.A, the subtraction of $r_1 = 1$ from d_1 involves borrow propagation and thus may affect all the bits of d_1. If we incorporate this subtraction in the next step where we perform the division by τ^2, then by putting $d_1 - 1$ in place of d_1 in Eq. A.1, we have the following observation.

$$
\begin{aligned}
(d_0, d_1) &\leftarrow \left(-\frac{2(d_1 - 1) + d_0}{4}, \ -\frac{2(d_1 - 1) - d_0}{4} \right) \\
&\leftarrow \left(-\frac{2d_1 + (d_0 - 2)}{4}, \ -\frac{2d_1 - (d_0 + 2)}{4} \right)
\end{aligned}
\tag{A.6}
$$

Thus we find that the subtraction of r_1 from d_1 is equivalent to the addition or subtraction of two with d_0. As $d_0 \equiv 2 \ (mod \ 8)$, the subtraction or addition of 2 changes only the three low bits of d_0.

In Case 5.C, the subtraction of $r_1 = -1$ from d_1 involves carry propagation. When we put $d_1 + 1$ in place of d_1 in Eq. A.1, we have the following observation.

$$
\begin{aligned}
(d_0, d_1) &\leftarrow \left(-\frac{2(d_1 + 1) + d_0}{4}, \ -\frac{2(d_1 + 1) - d_0}{4} \right) \\
&\leftarrow \left(\frac{2\bar{d_1} - d_0}{4}, \ \frac{2\bar{d_1} + d_0}{4} \right)
\end{aligned}
\tag{A.7}
$$

Algorithm 14: *New τNAF Generation Algorithm*

Input: Reduced Scalar $\gamma = d_0 + \tau d_1$
Output: $\tau\text{NAF}(\gamma)$

1 **begin**
2      $S \leftarrow \langle\rangle$ /* Used to store τNAF */ ;
3      $Sign \leftarrow 0$ /* Used to keep sign of (d_0, d_1) */ ;
4   **while** $d_0 \neq 0$ *or* $d_1 \neq 0$ **do**
5         $(r_0, r_1) \leftarrow Table\ 2$;
6         $(d_0, d_1) \leftarrow Alter\,Low\,Bits(d_0, d_1)\ in\ Sect.\ A.1$;
7         $(B, C) \leftarrow (0, 0)$ /* Borrow and Carry Inputs */ ;
8   **if** $Sign = 1$ **then**
9 | $(r_0, r_1) \leftarrow (-r_0, -r_1)$;
10   **end**
11        $Prepend\ (r_1, r_0)\ to\ S$ /* τNAF digits */ ;
12   **if** $Case\ 4\ True$ **then**
13 | $(B, C) \leftarrow (1, 1)$
14   **end**
15   **if** $Case\ 5.C\ True$ **then**
16 | $(d_0, d_1) \leftarrow (\frac{2\bar{d_1} - d_0}{4}, \frac{2\bar{d_1} + d_0}{4})$;
17 | $Sign \leftarrow Sign$;
18   **end**
19   **else**
20 | $(d_0, d_1) \leftarrow (\frac{2d_1 + d_0 + C}{4}, \frac{2d_1 - d_0 - B}{4})$;
21 | $Sign \leftarrow 1 \oplus Sign$;
22   **end**
23   **end**
24 **end**

So, using one's complement of d_1 in Case 5.C, we eliminate the long subtraction of r_1 from d_1. Computing one's complement in hardware platform is easy as all the bits of d_1 can be altered in parallel.

Algorithm 21 describes the steps of the new τNAF generation technique. Only one addition or subtraction operations are performed on d_0 and d_1 during division by τ^2 in any iteration. Thus, for the τNAF generation part of the scalar conversion, presence of only one adder/subtracter circuits in the critical paths of d_0 and d_1 is sufficient.

A.2 Hardware Architecture

We use these optimizations to design a high-speed and pipelined scalar conversion architecture. The architecture is described in details our publication.

Sujoy Sinha Roy, Junfeng Fan, Ingrid Verbauwhede. Accelerating Scalar Conversion for Koblitz Curve Cryptoprocessors on Hardware Platforms In *IEEE Transactions on Very Large Scale Integration (VLSI) Systems, vol. 23, no. 5, pp. 810–818, May 2015.*

Table A.3 Performance results on Xilinx Virtex 4 FPGA

Work	Slices	Freq MHz	Reduction time (μs)	Conversion time (μs)
Brumley et al. [1]	1671	65.9	4.3	8.6
Adikari et al. [1]	1998	65.1	2.2	4.4
Our implementation	1814	107	1.3	2.6

We would like to mention that the high-speed architecture is a proof of concept implementation and is not resistant against side channel attacks.

A.3 Implementation Results

Table A.3 shows the performance of our high-speed scalar conversion architecture for K-283. Results were obtained from Xilinx ISEv12.2 tool after place and route analysis with optimization for speed. The conversion time is the total time required for the scalar reduction and the complete τNAF generation.

Appendix B
Implementation of Operations Used by Algorithm 6

The operations required by Algorithm 6 are implemented by combining an FSM and a program ROM. The program ROM includes subprograms for all operations of Algorithm 6 and the FSM sets the address of the ROM to the first instruction of the subprogram according the phase of the algorithm and t_{i+1}, t_i.

Table B.1 shows the contents of the program ROM. The operations required by Algorithm 6 are in this ROM as follows:

- Line 0 obtains the next bits of the zero-free representation.
- Lines 1–23 perform the precomputation that computes $(x_{+1}, y_{+1}) = \phi(P) + P$ and $(x_{-1}, y_{-1}) = \phi(P) - P$.
- Line 24 computes the negative of Q during the initialization and Line 25 is the corresponding dummy operation.
- Lines 26–28 randomize the projective coordinates of Q by using the random $r \in \mathbb{F}_{2^{283}}$ which is stored in Z.
- Lines 29–34 compute two Frobenius endomorphisms for Q.
- Lines 35–37 set $(x_p, y_p) \leftarrow (x_{+1}, y_{+1}) = \phi(P) + P$ and compute the y-coordinate of its negative to y_m.
- Lines 38–40 set $(x_p, y_p) \leftarrow (x_{-1}, y_{-1}) = \phi(P) - P$ and compute the y-coordinate of its negative to y_m.
- Lines 41–43 compute $(x_p, y_p) \leftarrow -(x_{+1}, y_{+1}) = -\phi(P) - P$ and set the y-coordinate of its negative to y_m.
- Lines 44–46 compute $(x_p, y_p) \leftarrow -(x_{-1}, y_{-1}) = -\phi(P) + P$ and set the y-coordinate of its negative to y_m.
- Lines 47–66 compute the point addition $(X, Y, Z) \leftarrow (X, Y, Z) + (x_p, y_p)$ in López-Dahab coordinates using the equations from [2].
- Lines 67–71 recover the affine coordinates of Q by computing $(X, Y) \leftarrow (X/Z, Y/Z^2)$.
- Lines 72–73 and lines 74–75 initialize Q with (x_{+1}, y_{+1}) and (x_{-1}, y_{-1}), respectively, and lines 76–77 perform a dummy operation for these operations.

© Springer Nature Singapore Pte Ltd. 2020
S. Sinha Roy and I. Verbauwhede, *Lattice-Based Public-Key Cryptography in Hardware*,
Computer Architecture and Design Methodologies,
https://doi.org/10.1007/978-981-32-9994-8

Table B.1 The program ROM includes instructions for the following operations

0	Convert(k)	20	$y_{-1} \leftarrow y_{-1} \times T_1$	40	$y_m \leftarrow x_{-1} + y_{-1}$	60	$T_2 \leftarrow x_p \times Z$
1	$x_{+1} \leftarrow X^2$	21	$y_{-1} \leftarrow y_{-1} + x_{-1}$	41	$x_p \leftarrow x_{+1}$	61	$T_2 \leftarrow T_2 + X$
2	$y_{+1} \leftarrow Y^2$	22	$y_{-1} \leftarrow y_{-1} + Y$	42	$y_m \leftarrow y_{+1}$	62	$Y \leftarrow Y + Z$
3	$x_{+1} \leftarrow X + x_{+1}$	23	$y_{-1} \leftarrow y_{-1} + X$	43	$y_p \leftarrow x_{+1} + y_{+1}$	63	$Y \leftarrow Y \times T_2$
4	$x_{-1} \leftarrow x_{+1}^{-1}$	24	$Y \leftarrow X + Y$	44	$x_p \leftarrow x_{-1}$	64	$T_1 \leftarrow Z^2$
5	$T_1 \leftarrow Y + y_{+1}$	25	$T_1 \leftarrow X + Y$	45	$y_m \leftarrow y_{-1}$	65	$T_1 \leftarrow T_1 \times y_m$
6	$y_{-1} \leftarrow T_1 \times x_{-1}$	26	$X \leftarrow X \times Z$	46	$y_p \leftarrow x_{-1} + y_{-1}$	66	$Y \leftarrow Y + T_1$
7	$T_1 \leftarrow y_{-1}^2$	27	$T_1 \leftarrow Z^2$	47	$T_1 \leftarrow Z^2$	67	$x_{+1} \leftarrow Z$
8	$T_1 \leftarrow T_1 + y_{-1}$	28	$Y \leftarrow Y \times T_1$	48	$T_1 \leftarrow T_1 \times y_p$	68	$x_{-1} \leftarrow x_{+1}^{-1}$
9	$x_{+1} \leftarrow T_1 + x_{+1}$	29	$Y \leftarrow Y^2$	49	$T_1 \leftarrow T_1 + Y$	69	$X \leftarrow X \times x_{-1}$
10	$T_1 \leftarrow x_{+1} + X$	30	$Y \leftarrow Y^2$	50	$T_2 \leftarrow Z \times x_p$	70	$x_{-1} \leftarrow x_{-1}^2$
11	$y_{+1} \leftarrow T_1 + y_{-1}$	31	$X \leftarrow X^2$	51	$T_2 \leftarrow T_2 + X$	71	$Y \leftarrow Y \times x_{-1}$
12	$y_{+1} \leftarrow y_{+1} + x_{+1}$	32	$X \leftarrow X^2$	52	$X \leftarrow T_2^2$	72	$X \leftarrow x_{+1}$
13	$y_{+1} \leftarrow y_{+1} + Y$	33	$Z \leftarrow Z^2$	53	$X \leftarrow X + T_1$	73	$Y \leftarrow y_{+1}$
14	$x_{-1} \leftarrow x_{-1} \times X$	34	$Z \leftarrow Z^2$	54	$T_2 \leftarrow T_2 \times Z$	74	$X \leftarrow x_{-1}$
15	$y_{-1} \leftarrow y_{-1} + x_{-1}$	35	$x_p \leftarrow x_{+1}$	55	$X \leftarrow X \times T_2$	75	$Y \leftarrow y_{-1}$
16	$T_1 \leftarrow x_{-1}^2$	36	$y_p \leftarrow y_{+1}$	56	$Y \leftarrow T_1 \times T_2$	76	$T_1 \leftarrow x_{+1}$
17	$x_{-1} \leftarrow x_{-1} + T_1$	37	$y_m \leftarrow x_{+1} + y_{+1}$	57	$T_1 \leftarrow T_1^2$	77	$T_2 \leftarrow y_{+1}$
18	$x_{-1} \leftarrow x_{-1} + x_{+1}$	38	$x_p \leftarrow x_{-1}$	58	$X \leftarrow X + T_1$		
19	$T_1 \leftarrow x_{-1} + X$	39	$y_p \leftarrow y_{-1}$	59	$Z \leftarrow T_2^2$		

Point addition and point subtraction are computed with exactly the same sequence of operations. This is achieved by introducing an initialization which sets the values of three internal variables x_p, y_p, and y_m according to Table B.2 (these are in lines 35–46 in Table B.1). This always requires two copy instructions followed by an addition. After this initialization, both point addition and point subtraction are computed with a common sequence of operations which adds the point (x_p, y_p) to Q. The element x_m is the y-coordinate of the negative of (x_p, y_p) and it is also used during the point addition.

Table B.2 Initialization of point addition and point subtraction

t_{i+1}, t_i	1st	2nd	3rd
$+1, +1$	$x_p \leftarrow x_{+1}$	$y_p \leftarrow y_{+1}$	$y_m \leftarrow x_{+1} + y_{+1}$
$+1, -1$	$x_p \leftarrow x_{-1}$	$y_p \leftarrow y_{-1}$	$y_m \leftarrow x_{-1} + y_{-1}$
$-1, +1$	$x_p \leftarrow x_{-1}$	$y_m \leftarrow y_{-1}$	$y_p \leftarrow x_{-1} + y_{-1}$
$-1, -1$	$x_p \leftarrow x_{+1}$	$y_m \leftarrow y_{+1}$	$y_p \leftarrow x_{+1} + y_{+1}$

Curriculum Vitae

Sujoy Sinha Roy was born on June 4th in Hooghly, India. He received the B.E degree in Electronics and Telecommunication Engineering from Bengal Engineering and Science University, Shibpur in 2007. He subsequently worked as an engineer in Tata Consultancy Services, Mumbai. In 2012, he received the M.S. degree in Computer Science and Engineering from Indian Institute of Technology, Kharagpur.

In September 2012, he started his PhD at the COSIC (Computer Security and Industrial Cryptography) research group at the Department of Electrical Engineering (ESAT) of the KU Leuven. His research area has been broadly in the field of efficient implementation of public key cryptography. His research has been generously funded by the European Commission through the Erasmus Mundus PhD Scholarship.

© Springer Nature Singapore Pte Ltd. 2020   99
S. Sinha Roy and I. Verbauwhede, *Lattice-Based Public-Key Cryptography in Hardware*,
Computer Architecture and Design Methodologies,
https://doi.org/10.1007/978-981-32-9994-8

References

1. Adikari J, Dimitrov VS, Järvinen K (2012) A fast hardware architecture for integer to τNAF conversion for Koblitz curves. IEEE Trans Comput 61(5):732–737 May
2. Al-Daoud E, Mahmod R, Rushdan M, Kilicman A (2002) A new addition formula for Elliptic curves over $GF(2^n)$. IEEE Trans Comput 51(8):972–975 Aug.

© Springer Nature Singapore Pte Ltd. 2020
S. Sinha Roy and I. Verbauwhede, *Lattice-Based Public-Key Cryptography in Hardware*,
Computer Architecture and Design Methodologies,
https://doi.org/10.1007/978-981-32-9994-8

Printed in the United States
By Bookmasters

Printed in the United States
by Baker & Taylor Publisher Services